CHASING
SHACKLETON

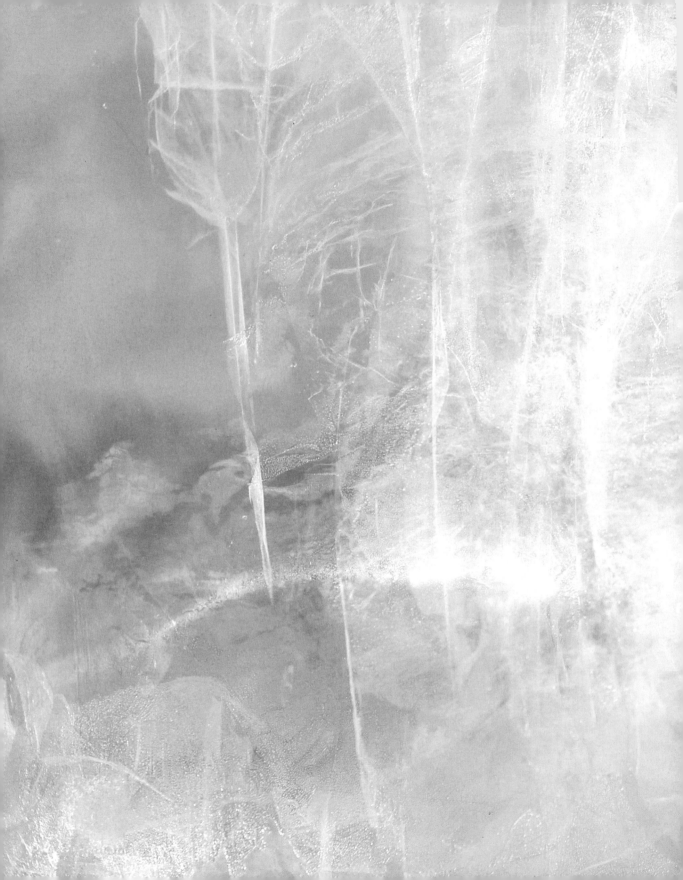

CHASING
SHACKLETON

RE-CREATING THE WORLD'S
GREATEST
JOURNEY OF
SURVIVAL

TIM JARVIS

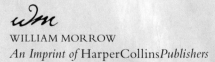

WILLIAM MORROW

An Imprint of HarperCollinsPublishers

TO WILLIAM AND JACK

First published as *Shackleton's Epic* in Australia in 2013 by HarperCollins Publishers Australia Pty Limited.

HarperCollins books may be purchased for educational, business, or sales promotional use. For information please e-mail the Special Markets Department at SPsales@harpercollins.com.

FIRST U.S. EDITION

Designed by Matt Stanton

Library of Congress Cataloging-in-Publication Data has been applied for.

ISBN 978-0-06-228273-6

14   15   16   17   18   [QGT]   10   9   8   7   6   5   4   3   2   1

CONTENTS

FOREWORD

This is a story of triumph! Tim Jarvis and his companions of the Shackleton Epic Expedition (SEE) have successfully re-created my grandfather Ernest Shackleton's 1916 voyage in the tiny *James Caird* (6.9 meters long, or not quite 23 feet). Ernest Shackleton crossed the 800 nautical miles from Elephant Island to South Georgia over the stormiest seas in the world and subsequently climbed the mountainous, unmapped interior of South Georgia.

Shackleton's aim was, of course, to rescue his twenty-two men marooned on Elephant Island. Tim's aim was to pay tribute to Ernest Shackleton's leadership, of which the 1916 expedition is regarded as the finest example. Tim is a veteran of sixteen expeditions and when I met him I had no doubt that he was the man who could make the SEE happen. The team was chosen and finally, after years of preparation, the replica boat was built and named *Alexandra Shackleton* after myself as patron, a great honor. When she was lowered into the sea at Portland Marina, she looked small but resolute. The SEE set off on their great adventure early in 2013.

How Shackletonian was the Shackleton expedition?

About one hundred years ago, Ernest Shackleton listed the qualities he required in a polar explorer.

First, he listed optimism. During the five years it took to bring the SEE into being, Tim never wavered in his belief that the expedition would become a reality.

Second, Ernest Shackleton listed patience. Tim required a great deal of patience for long, drawn-out negotiations with government departments, sponsors, supporters, and the media. Sorting out the logistics and sourcing the equipment was also a major operation, and a lot of work had to be done on the *Alexandra*

Shackleton. A considerable contribution was made by the other members of the expedition, Nick Bubb, Baz Gray, Paul Larsen, Seb Coulthard, and Ed Wardle.

Third, Ernest Shackleton listed imagination, with which he coupled idealism. Tim had the imagination to see that the expedition's use of original-type clothing, original-type equipment, and original-type food would bring them physically as well as spiritually closer to the spirit of Ernest Shackleton. And Tim had the idealism to hold to his vision, building a legend on a legend.

Finally, Ernest Shackleton listed courage. It takes courage to embark in such a tiny boat in such seas. If any of the expedition had fallen overboard, they probably would not have survived. And although the dangerous climb over South Georgia's mountains was delayed by eighty-knot winds, there was no question of not carrying on.

So, all in all, I feel that the SEE was a thoroughly Shackletonian expedition.

As for myself, the high winds delayed my ship's entry into Grytviken Harbor. It was an anxious time. I knew that all the SEE were well and what they had achieved, but our planned rendezvous at my grandfather's grave was very important to us all. However, the wind dropped and South Georgia produced its best blue-and-gold day, so bright it was difficult to see. I was reunited with Tim and the team and, as tradition demands, we drank a toast to Ernest Shackleton and poured a libation over his grave.

It was a great moment for a very proud patron.

Alexandra Shackleton

SOUTH 1

The trails of the world be countless,
and most of the trails be tried;
You tread on the heels of the many,
till you come where the ways divide;
And one lies safe in the sunlight,
and the other is dreary and wan,
Yet you look aslant at the Lone Trail,
and the Lone Trail lures you on.

Robert Service, *The Lone Trail*

I thought I knew Antarctica by now. I had been to its frozen alien shores, a world with no native human population, three times. I had become, to the extent that one can, "used" to the highest, coldest, windiest continent in the world with its extreme weather and the staggering, kilometers-thick mantle of ice that covers it.

My initial expedition into the polar regions had been a trek of tortuous slowness across the island of Spitsbergen in the high Arctic with my close friend Andrew "Ed" Edwards, where the danger of polar bear attacks and crevasses challenged us to our limits and revealed a strength and determination I wasn't aware I possessed. In 1999 I'd taken on what many regard as one of the last great land-based challenges on earth—crossing the continent's 2,700 kilometers on foot and unsupported, pulling a sled weighing 225 kilograms through obstructive icy terrain. Among other consequences, I'd seen my fingers blackened by frostbite; experienced temperatures so low that three of my metal fillings contracted and fell out, requiring self-administered dental repairs; lost 20 percent of my body weight; eaten a sickness-inducing 7,200 calories of lard and olive oil each day; and written "That was the toughest day of my life" in my diary on seventeen consecutive days. On that occasion, my journey ended early, when a ruptured fuel container resulted in food contamination. Nevertheless, I had covered 1,800 kilometers and reached the Pole in a record forty-seven days, allowing even someone as self-critical as me to be rightly proud of what had been achieved.

Fate played its hand in my next journey, which was south to the Antarctic. For my work as a scientist I had moved to Adelaide in South Australia. This

In times of trouble pray God for Shackleton.

Previous pages: Point Wild, the place Shackleton's twenty-two men would call home for four months, complete with characteristic brash ice.

brought me into unlikely contact with the legacy of Australia's greatest land-based polar explorer and an Adelaide legend, Sir Douglas Mawson.

In 1913 Mawson was forced to undertake an incredible survival journey. While mapping an uncharted section of the Antarctic coast as part of the Australasian Antarctic Expedition, he lost the first of his two companions, Belgrave Ninnis, and the dog sled that contained most of the expedition's food and equipment in a crevasse fall. What followed was starvation, blizzards, debilitating cold, and, ultimately, following the consumption of the remaining sled dogs, the death, in Mawson's arms, of his second companion, Xavier Mertz, of what he described at the time as "fever." Alone, Mawson faced ferocious winds, near-fatal crevasse falls, and terrible debilitation, all compounded by the loneliness and danger of solo travel. When, against all odds, he finally stumbled through the door of his hut fifty days later, his men asked, "Which one are you?" Mawson's shocking physical state made him unrecognizable. With some having accused Mawson of cannibalizing Mertz in order to survive, I decided I would re-enact the journey with what he said he had available to him, not only to test myself but also to see if I could shed light on Mawson's survival. When I returned to civilization, journey complete, I was asked for a word that described the hardship of surviving on my own on starvation rations in a frozen, reindeer-skin sleeping bag following the "death" of my colleague. All I could think of was "desperate."

But this time I was planning a very different journey. In attempting to re-create Sir Ernest Shackleton's legendary Antarctic survival trek across sea and ice in 1916, I would trade pulling a sled through mountains toward an endless white horizon for sailing and rowing a tiny, unstable wooden boat toward an endless gray one. Antarctica would be my starting point rather than my final destination. And I would be on a journey where the Antarctic weather that raged all around us would not only threaten from above but also turn the ocean across which we traveled into a tortured, ever-changing landscape of terrifying proportions.

The prospect of what lay ahead haunted me. Try as I might, I could not shake the image of a man in the dark water facing certain death, alone, watching his boat drift into the distance as the merciless cold of the Southern Ocean drained his lifeblood. Many thought the trip was virtually impossible. As he set off in his tiny, keel-less boat to try to cross the Southern Ocean from Elephant Island to South Georgia, Shackleton had said to his skipper, Frank Worsley, "Do you know I know nothing about boat sailing?" Worsley assured him that, luckily, he did. Shackleton was as usual being self-effacing about his ability. I, on the other hand, was not: I knew very little about boat sailing and in my darkest moments it weighed heavily on me.

"It was an obsession that claimed them all," the curator whispered in revered tones. The "them" to which he referred were Scott, Shackleton, and Amundsen, their obsession the exploration of the polar regions during the heroic era of exploration in the early years of the twentieth century. Looking at the equipment they used, it seemed hardly surprising and made my attempt on the North Pole the following year in Gore-Tex and Kevlar seem somehow lightweight—both literally and metaphorically—compared to their sepia-hued, superhuman feats featured on the walls and in the display cabinets all around us.

"May I introduce Alexandra Shackleton, granddaughter of Sir Ernest?" said another voice beside me. This time it was that of my good friend Geraldine. I turned to greet Alexandra with the respect the Shackleton name instantly commands, particularly in the hallowed surrounds of the Greenwich Maritime Museum. It was 2002 and we were there for the opening of the exhibition *South*, a celebration of the achievements of Alexandra's grandfather, Scott, and Amundsen, but perhaps also a recognition of the esteem with which Shackleton's account of the *Endurance* expedition of 1914–17 of the same name was regarded.

Left: Mawson—scientist, explorer, survivor.

Right: Me, re-creating Mawson's desperate journey of survival.

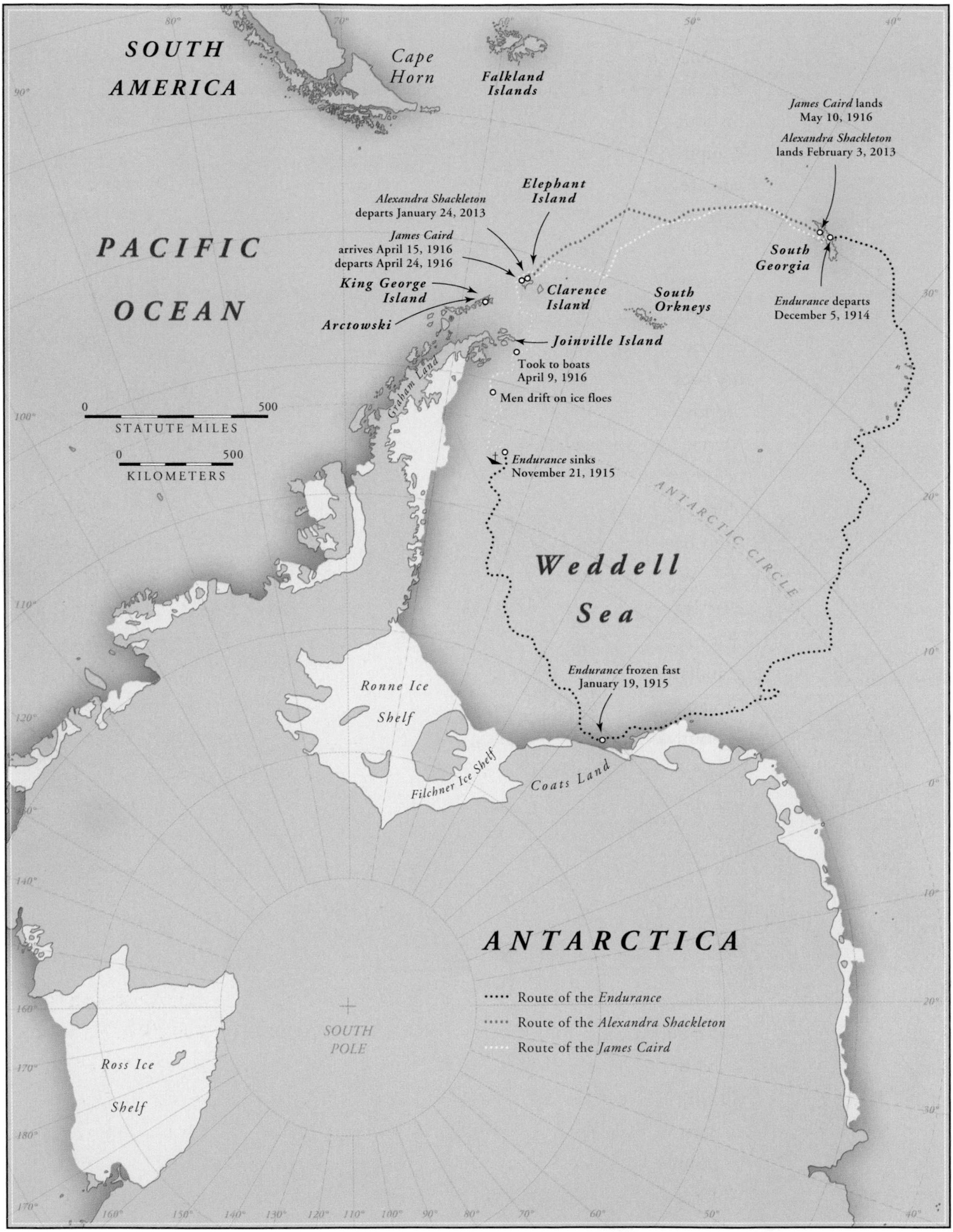

SOUTH
AMERICA

Cape
Horn

Falkland
Islands

James Caird lands
May 10, 1916

Alexandra Shackleton
lands February 3, 2013

PACIFIC

Alexandra Shackleton
departs January 24, 2013

James Caird
arrives April 15, 1916
departs April 24, 1916

**Elephant
Island**

OCEAN

**King George
Island**

Arctowski

*Clarence
Island*

*South
Orkneys*

**South
Georgia**

Endurance departs
December 5, 1914

Joinville Island

Took to boats
April 9, 1916

Men drift on ice floes

ANTARCTIC CIRCLE

200
STATUTE MILES

500

0   500
KILOMETERS

Endurance sinks
November 21, 1915

Graham Land

W e d d e l l

S e a

Endurance frozen fast
January 19, 1915

*Ronne Ice
Shelf*

Filchner Ice Shelf

Coats Land

A N T A R C T I C A

SOUTH
POLE

*Ross Ice
Shelf*

····· Route of the *Endurance*
····· Route of the *Alexandra Shackleton*
····· Route of the *James Caird*

Mawson's achievements were noticeably absent from the exhibition, but Zaz, as Alexandra prefers to be known, was intrigued by my plans to retrace his journey the old way with the same starvation rations and hundred-year-old equipment. "It sounds fascinating," she commented. "And what might you do if you are successful with that journey?" The significance of this question would not become clear until years later.

On my completion of the Mawson expedition, Zaz was one of the first to call to congratulate me on my success and praise the way in which I'd done it. I had kept it as true to the original journey as possible, with the notable exceptions being that no one died and we ate neither dogs nor men. This was something of a relief for my backers but even more so for my expedition partner, John Stoukalo, who was slightly concerned at the prospect of having to die halfway through like the ill-fated Mertz. The trip had been incredibly challenging, with more weight loss than ever before, a return of the old frostbite injuries plus a few new ones, and the need to plumb new depths of physical and mental resolve in order to complete the journey. But I had seen no need for the calories that eating another would have provided.

"What next?" Zaz asked innocently enough but with both of us knowing exactly what she meant. Through our close friendship that had developed since our first meeting, I knew she rued the fact that no one had successfully re-created her grandfather's famous "double" as he had done it—a journey across the Southern Ocean in a replica *James Caird* followed by a climb across the mountainous interior of South Georgia. When one looked at the difficulty levels and the inherent danger, it was hardly surprising. "I would like you to lead a team to attempt this," she stated. They were powerful words and, although I had anticipated them, they still made my pulse quicken. "I would be proud to," I replied. With those few words I knew a cast-iron commitment had been made, one that Shackleton would have expected me to honor and that neither of us would let go.

The route of the ill-fated Imperial Trans-Antarctic Expedition, 1914–1916.

Shackleton's original expedition followed Amundsen and Scott, reaching the South Pole in 1912. Not to be outdone, Shackleton decided to embark on the most ambitious polar expedition of them all—the Imperial Trans-Antarctic Expedition (ITAE), a bid to cross Antarctica on foot from the Weddell Sea coast to the Ross Sea coast in what he described as "the one great main object of Antarctic journeyings." In an interview for the *Daily Mirror* entitled "My Talk with Sir Ernest Shackleton," William Pollock asked Shackleton why he was going on a South Polar expedition after Amundsen and Scott had succeeded in reaching the Pole itself. "He began to talk of the scientific, geographical and other benefits which he hoped would result from such an expedition," wrote Pollock, "and then, suddenly fixing his eyes upon me, he said: 'Besides, there's a peculiar fascination about going. It's hard to explain it in words—I don't think I can quite explain it—but there's an excitement, a thrill—a sort of magnetic attraction about polar exploration.'"

ITAE planned to use two ships to accomplish its goal. The first ship, the *Endurance*, on which Shackleton traveled, would land at a site near Vahsel Bay, adjacent to the Ronne Ice Shelf in the Weddell Sea. From here Shackleton would begin his attempt to cross the continent by a route that interestingly was very similar to the starting point of my bid to cross Antarctica in 1999–2000 that left from nearby Berkner Island on the Ronne Ice Shelf. The second ship, Mawson's former vessel the *Aurora*, would leave from Hobart under the command of Aeneas Mackintosh and land at McMurdo Sound on the Ross Sea side. Its men would then lay a series of food caches in toward the Pole from their side that the crossing team would access once they passed the Pole.

Shackleton had learned from mistakes made on previous expeditions and was taking a large team of dogs, dietary precautions against scurvy, and a Royal Marine physical-fitness instructor, Thomas Orde-Lees, whose role among other things would be to teach the men to ski. Their improved diet, the result of painstaking research and analysis by Shackleton and Colonel Wilfred Beveridge of the Royal Army Medical Corps in a bid to minimize the risk of scurvy, undoubtedly helped their cause. It turned out, however, that neither the dogs nor an ability to ski would be needed, given the events that transpired.

The *Endurance* left Grytviken, South Georgia, in early December 1914 and headed south, bound for Vahsel Bay, in a year when the sea ice was the worst the whalers had ever experienced. For a week the ship, which was powered by engine and sail, barged and cajoled her way through the pack, her thick hull specifically designed for the purpose. But with Vahsel Bay still some 135 kilometers distant, the ice finally formed an impenetrable barrier many meters thick to the horizon in every direction. The same winds that supplemented the power from the *Endurance*'s engines by filling her sails and pushing her onward were, ironically, largely responsible for driving the vast mass of pack ice hard up against Antarctica, trapping them in the process.

After many attempts to free themselves, Shackleton announced on February 24 that the ship was officially a winter station and suspended ship routine, accepting

What the ice gets, the ice does not surrender: the Endurance *beset by ice.*

that they were not going to escape the ice until the following spring or summer. He now had to get twenty-eight men from disparate backgrounds to live together harmoniously—not easy given that the sailors had been expecting to head back to civilization soon after dropping off the "shore party" of expeditioners and scientists. With big personalities involved and wide-ranging personal likes and dislikes bubbling below the surface, it was a huge challenge.

Shackleton established a structured routine of social activity, including lantern evenings, regular exercise, and tending to the dogs, and he relocated all of the men's living quarters down into the warmest part of the ship. Now the eccentricities of his recruitment process came to the fore: the optimism and flexibility he had looked for in each man began to pay dividends. Shackleton held optimism almost above all else, calling it "true moral courage," and they would need all they had to get through.

The *Endurance* remained beset until September, when the ice started to break up. The men greeted this positively and started speculating about their being freed and perhaps being able to continue south. But actually it signified great

danger—the kind of danger one gets when rafts of ice many meters thick and the size of cities are driven together by powerful forces of wind and currents. The resulting "pressure" will crush anything in its path, even the strongest ice-strengthened vessel like the *Endurance*, especially when she was embedded in the ice. "Pressure" was a very apt description of the situation in which they now found themselves: on their own in this alien world with no one knowing they were there and with no means of communicating with anyone.

By October the intense pressure of the ice had breached the stricken ship's hull and she was sinking despite bilge pumps and men operating around the clock to try to save her. On October 27 Shackleton ordered the men to abandon ship, setting up camp in tents on the ice nearby. Immediately and with typical decisiveness, he determined that they would prepare to march toward the tip of the Antarctic Peninsula, some 400 kilometers to the northwest. Shackleton's ability to refocus on new goals and his characteristic optimism and conviction were clearly demonstrated by his calm announcement to the men: "So now we'll go home."

Their bullishness was soon dampened, however, as they discovered the impossibility of pulling the lifeboats across the contorted surface of pack ice. The three lifeboats—the *James Caird*, *Dudley Docker*, and *Stancomb Wills*, named after the expedition's sponsors—had been rescued from the *Endurance* and would be their only way home. But each boat weighed more than a tonne (a metric ton) and, despite being on sleds, was desperately heavy and cumbersome to pull. I can certainly attest to the difficulty of pulling a sled through the pack ice of the Arctic Ocean—like a building site with walls and piles of frozen rubble many meters high, over and through which you need to pick your way. A more demoralizing and confused surface would be difficult to imagine.

In light of the circumstances, Shackleton changed his plan and decided to set up Ocean Camp less than three kilometers from the wreck of the *Endurance*. The goal now was to hope they drifted northwest in the pack so that when it ultimately broke up, they would be free to complete the remainder of the journey at sea in the boats, sticking close to shore. It was a tense time as the wind appeared not to have read the script, sending them backward and out to sea as often as toward land. Meanwhile, Shackleton battled with severe sciatica and the men suffered in damp sleeping bags, the wood salvaged from the *Endurance* not insulating them sufficiently from the snow and ice beneath. They were also worried they would run out of food, given their rate of consumption. Just when any sane consideration of their circumstances would surely have resulted in feelings of utter hopelessness

A man's best friends: Shackleton's ship and his dogs on ice.

and despondency, Shackleton's optimism again came to the fore. Such an ability to look favorably at one's predicament, almost to the point of self-delusion in the face of the awful truth of one's circumstances, is crucial to every polar expeditioner, especially given the enormity of the task one sets oneself in places where the chances of success are low and problems and doubts unrelenting. It was a skill over which Shackleton had complete mastery, but it was not to everyone's liking. The *Endurance*'s first officer, Lionel Greenstreet, referred to it as "absolute foolishness," summing up what quite a few of the men thought.

It was here at Ocean Camp that Shackleton got the expedition's carpenter, Henry "Chippy" McNeish, to begin preparing the *Caird* and the other boats for a long sea journey. With extremely limited resources, McNeish managed to add thirty-five centimeters to the gunwales of the *Caird* using nails from the *Endurance* and filling the seams with lamp wick and the oil paints of Marston, the artist. This gave the *Caird* some seventy centimeters of freeboard, all of which would be needed for the journey ahead.

Despite Chippy McNeish's excellent work, relations were not good between

Man-hauling the boats: in a desperate bid to reach the open sea and be free of the ice, the crew tried to drag the boats by hand.

him and Shackleton. The carpenter, who feared the drift was carrying them out to sea, had vocally disagreed with Shackleton's latest decision to begin marching again toward land. The rift between the two men would never heal, as Shackleton felt McNeish's behavior was tantamount to mutiny. "I shall never forget him in this time of stress," he lamented. In the end, ice conditions forced a rethink anyway and a move to stronger ice nearby and what they dubbed "Patience Camp." Here, a more candid assessment of their dire food situation resulted in the need to shoot twenty-seven of their dogs that they could no longer afford to feed, Frank Wild reporting that it was the worst job he had ever had to do: "I have known many men I would rather shoot than the worst of the dogs."

Fear and uncertainty stalked the camp as the men drifted north, parallel with the peninsula and tantalizingly close to Paulet and Joinville islands, less than five kilometers to the west and the last islands before the end of land, the vastness of the Southern Ocean, and an even worse fate.

Over the course of the next week, the men were carried farther north out into the open ocean by the strong currents, their soccer-field-sized home rapidly

disintegrating beneath them in the big ocean swell—a two-meter-thick leaf on a 2,000-meter-deep pond. Finally, a further crack in their floe forced their hand and they launched the boats on April 9, knowing that either Clarence or Elephant Island, the mountains of which had appeared on the horizon, was their last chance of survival. If they missed the islands, certain death at sea awaited. It was the austral autumn of 1916, the First World War raged on, and the crew of the *Endurance* were twenty-eight men in three small wooden lifeboats adrift in the roughest ocean in the world.

Their journey was a terrifying one. Initially it involved trying to follow leads in the shifting pack ice that dangerously opened and closed with a force that could crush the boats in an instant. Although terribly dangerous, at least the pack afforded protection from the open water that was far rougher without the dampening effect of the ice. Plus each night they could at least camp on a suitable floe—a better option than remaining on the open sea.

Based on the constantly changing winds, at one stage Shackleton opted to aim for King George Island to the west. Then the winds changed again, driving them depressingly back beyond Patience Camp. At this stage they had no drinking water left and the men were exhausted and understandably fearful of what might happen next. Temperatures were well below freezing, snow was falling, waves were crashing into the boats, and the *Stancomb Wills*, whose gunwales had not been raised, was awash with knee-deep, freezing seawater. Hypothermia was close at hand and the men were suffering from trench foot—an ailment not unlike frostbite—caused by the cold, damp, restricted conditions. In addition, a dangerous apathy was setting in, born in roughly equal parts of their being completely at the mercy of the elements and their utter exhaustion. Shackleton decided to head for Elephant Island, admitting privately that he "doubted if all the men would survive that night." Elephant Island was deemed to be the better option largely because the winds they had would allow an attempt at Clarence if they missed it. If they missed Clarence Island, that would be it: there was no more land save South Georgia, an impossible 800 nautical miles to the northeast—an inhospitable dot in a very large ocean.

Shackleton approached the underbelly of Elephant Island in a howling gale, aiming for a broad bay some seventeen miles wide on its southeastern side that they could not see due to the "blackness of the gale and thick snow squalls," according to Worsley. They approached cautiously until the wind yet again changed direction, blowing from the southwest, directly behind them, causing

On Elephant Island, the end of an improbable journey for two lifeboats—but only the beginning for the James Caird.

heavy, confused seas. Worsley, who was skippering the *Dudley Docker*, felt there was serious danger of capsizing. Finally, they managed to round Cape Valentine, the strength of the sea abating and the gale decreasing as they moved into the lee of the island, but it had been a desperately close run.

Sealers named Elephant Island, or Sea Elephant Island, after the massive creatures that lived there, the only mammals that had managed to establish themselves on its inhospitable shores. Even the sealers and whalers themselves had been unable to do so due to its remoteness, rough weather, and absence of a sheltered anchorage. Every indentation of Elephant Island's rugged coastline is steep glacial ice save for the dark rock cliffs that descend directly into an ocean of huge seas crashing unrelentingly at their base.

After their arrival, the mist cleared, as expedition photographer Frank Hurley's description of Cape Valentine attests. "Such a wild and inhospitable coast I have never beheld," he wrote, going on to describe "the vast headland, black and menacing, that rose from a seething surf 1,200 feet above our heads and so sheer as to have the appearance of overhanging." All in all this meant that, other than representing terra firma, Cape Valentine was a very poor prospect for their ongoing survival. Their position on a narrow, shingle beach underneath overhanging cliffs meant it was a race between a big sea washing them away or snow, ice, and rocks cascading down on them from the cliffs above. They had to move immediately.

Frank Wild again took to the sea in the *Dudley Docker* to find a better camp,

and chose the spot later named Point Wild by the men in his honor. The name could just as easily have been a description of the site itself—a shingle and rock spit extending perpendicular into the sea capped by large, rocky outcrops on the seaward side. Seas from both east and west battered it. The western side, meanwhile, was routinely choked with ice from the glacier only 200 meters away, major ice falls created enormous waves that on several occasions almost engulfed Shackleton's puny camp on the shingle beach. The violent, contorted river of glacial ice prevented progress anywhere on foot to the south and west while 300-meter-high black cliffs did the same to the east.

All were relieved to have made it to this inhospitable, rocky outcrop, no one more so than Shackleton. But his relief was soon overshadowed by the knowledge that, on an island without a human population and seldom visited even by whalers, they would never be found. He knew, too, that only a thin veneer of morale remained among many of the men despite the joy and relief—initially at least—at having made it this far, and that neither morale nor men would likely survive the fast-approaching winter. It was April 1916, and with autumn upon them already, he knew that he would have to bring civilization to them, and waste no time in doing it.

The journey that followed unbelievably eclipsed what had gone before and was described by Sir Edmund Hillary as "the greatest survival journey of all time." Shackleton and five of his most able men left Elephant Island on April 24 on an 800-nautical-mile voyage across the notoriously treacherous Southern Ocean in the largest of the three lifeboats, the *James Caird*.

Frank Worsley was skipper. The New Zealander was something of an eccentric, a risk taker and not a natural leader—in the authoritarian sense at least—but he was undoubtedly one of the most accomplished sailors of his time. Tom Crean was a tough Irishman from County Kerry and perhaps the strongest polar expeditioner of the heroic era. He was fiercely loyal to Shackleton and virtually begged to be included in the *Caird* team. His experience and resilience were invaluable: as a petty officer in the Royal Navy, Crean had already served under Scott on two of his Antarctic expeditions. Chippy McNeish, with whom Worsley in particular did not get on, was a rough Scot whom Shackleton described as "the only man I'm not dead certain of." But his efforts to make the *Caird* more seaworthy were remarkable and, as it turns out, critical. One of the oldest men on the *Endurance*, McNeish had a rough, cantankerous manner that was problematic, and Shackleton undoubtedly chose him for the *Caird* crew as

Iron Men: Shackleton's chosen few, clockwise from top left: Tom Crean, John Vincent, Chippy McNeish, Timothy McCarthy, and Frank Worsley.

much to safeguard morale among those left behind on Elephant Island as for his respected skills as a sailor and shipwright. Similarly, Shackleton chose John Vincent, a hardened ex-naval North Sea trawlerman, because of his physical strength and sailing skill, but in no small part to remove him from Elephant Island due to his bullying ways. Timothy McCarthy, meanwhile, was perhaps the best and most efficient of the sailors, always cheerful under the most trying circumstances and described by Worsley as "the most irrepressible optimist I've ever met." Having survived this great journey, he was tragically still claimed by the sea, dying as he did only three weeks after returning home on a navy ship that was lost with all hands during the Great War. It was McCarthy's first day under enemy fire.

For seventeen days this tough group of men battled constant gales, terrible cold, and mountainous seas in their twenty-three-foot keel-less wooden boat, using only Worsley's occasional sextant sightings from the boat's pitching deck to navigate by. That they not only found South Georgia but also managed to land on this remote island is incredible. Their epic voyage and subsequent survival is a remarkable testament to both Shackleton's leadership and the seamanship of Worsley, who saw the sun for only four sightings during the whole voyage in tumultuous seas.

Upon their arrival at South Georgia, following a hurricane that nearly finished them off, the six men clawed their way into King Haakon Bay, a spectacular fjord on the southwestern side of the island. There, Shackleton left Vincent and McNeish, who were too exhausted to continue, in the care of McCarthy. He, Worsley, and Crean then climbed over the unexplored, heavily glaciated mountains of South Georgia to reach Stromness whaling station on the other side. It was a thirty-five-kilometer journey that the world's top mountaineers in the modern era have subsequently been unable to replicate in the way Shackleton did it.

Ultimately Shackleton was able to save all of his men—the three who remained on the other side of South Georgia and, with the help of the Chilean Navy, all twenty-two of the crew members who had been left stranded on Elephant Island. It was an epic triumph of endurance and leadership.

This boat journey from Elephant Island to South Georgia and the climb across the mountains to Stromness was what Zaz had been referring to when she asked me to lead a team to attempt "the double." My unequivocal agreement meant I needed not only to build a replica *Caird* from scratch but also to assemble a team skilled and brave enough to attempt this with me.

So why did I do it? A simple answer was that I was honored to be asked by Shackleton's granddaughter to undertake this journey and was inspired to want to do it as the greatest survival story of the heroic era of exploration. Despite my reprising the conversation her grandfather and Worsley had had in which Shackleton suggested he was "no small-boat sailor," she remained unmoved and I am grateful to her for her confidence. In no small part, too, it felt like the logical conclusion of my bid to cross Antarctica in 1999, thwarted as it was by a fuel leak 1,800 kilometers into the crossing. I hadn't been trapped in the pack ice as Shackleton had been, unable even to land on Antarctica, but I still had unfinished business in the Great White South. Perhaps the journey in the replica *Caird* and climbing across South Georgia would be a closing chapter for us both.

At a more philosophical level, I consider exploration to be the adventure of seeing whether or not you can achieve something, the thrill of trying, and the process of learning more about yourself and your surroundings that going on a journey to find out affords you, and I think this outlook is consistent with Shackleton's. That we as individuals need to challenge ourselves to find out more about the world and our place in it is, I believe, as relevant a concept now as it was for Shackleton. As André Gide said so eloquently, "It is only in adventure that some people succeed in knowing themselves—in finding themselves."

Certainly this love of exploration both literal and personal was a major driver for Shackleton, as undoubtedly was the desire to excel, with polar exploration being the means by which this could happen. As a middle-class, merchant-navy man of Anglo-Irish parentage (despite moving to England at the age of ten, his voice apparently retained traces of his Irish roots), he did not naturally fit easily into either Irish or English society, obsessed as they were in Edwardian times—and to an extent still are—with social standing. As the outsider—unlike Scott, who in every way represented the establishment—Shackleton defied pigeonholing and actively resisted it, displaying a healthy disregard for class and tradition. Polar expeditions offered him a way to transcend these boundaries while fueling his love of adventure—and not hurting his marriage prospects either.

Perhaps it was this refreshingly modern attitude that resonated so strongly in my own mind, as much as his extraordinary achievements. The more I discovered about Shackleton, the more I recognized, from his adaptability born of living in a number of different places and needing to fit in, his determination to follow projects through to completion and the energy with which he did so, right down to the stubbornness and impatience he exhibited once he had decided upon a course of action. His intolerance of any negativity expressed by his men, along with his not being fazed by problems as long as a solution was suggested, were also familiar attitudes, not to mention his hunger for adventure and new experiences. In short, had we met over a few drinks, I think Shackleton and I would have had quite a few things to talk about and agree on.

And, of course, there were the many intriguing anecdotal details that made Shackleton such a compelling character and the story of his survival so remarkable. His powerful personal charm and charisma, which enabled him to talk as easily to Kaiser Wilhelm II as to the lowliest member of his crew and which won over financial backers for his expensive expeditions and gained him the unswerving loyalty of those who served under him (who affectionately called him "the Boss"), are the stuff of legend. Combine this with the failings he displayed both financially and personally and, for me, he became over time the most extraordinary and most "human" of all the heroic-era explorers. The invitation to retrace his journey was more than just an opportunity for adventure.

In the years since the achievements of Shackleton and his peers, there has been an ebb and flow in the way in which their exploits have been regarded. Mawson survived to old age, going on to achieve great things beyond his Antarctic feats, and is now finally beginning to get the recognition he deserves. Scott, once the hero who had paid the ultimate price by selflessly giving his life in the pursuit of his goal, has come to be seen by many as someone who went too far. Perhaps he was too readily prepared to sacrifice not only his life but also those of his men, and made serious mistakes along the way that contributed to their demise. Amundsen, who was seen as unfeeling and dispassionate with his routine butchering of his dogs and clinical way of operating on the ice, with time has revealed human failings. His poor judgment in initially leaving too early in his first bid for the South Pole meant his men barely escaped with their lives. One, Hjalmar Johansen, felt so betrayed he fell out with Amundsen and was dropped from the successful polar team. The shame he felt resulted in Johansen taking his own life.

For Scott, Shackleton, and Amundsen, *pro patria mori* ended up reigning supreme. Scott, the last to die in his tent, was writing eloquently until the end. Amundsen, having conquered the Northwest Passage by boat and the South Pole on foot, was lost in a plane searching for a former colleague, Umberto Nobile, in the Arctic. And as for Shackleton, he died only five years after the legendary journey he undertook after the loss of the *Endurance*.

Shackleton's star has risen and continues to do so, his decisiveness, compassion, and ability demonstrated so ably by his salvaging victory from the jaws of defeat in saving his men, representing an ideal to aim for in a world where selfless heroic leadership is aspired to by many, practiced by few, and needed by all. Now I was committed to emulating this most difficult of journeys by this most incredible of men and hoped that I could rise to the challenge. An apocryphal advertisement for the original expedition read, "Men wanted for hazardous journey. Safe return doubtful . . . honour and recognition in case of success."

Together alone: the Endurance *crew beside their ice-bound ship.*

THE ICE AGE

The heroic era of exploration—to which Ernest Shackleton and his contemporaries Robert Falcon Scott, Roald Amundsen, and Douglas Mawson belonged—began with the first Antarctic landing by Carsten Borchgrevink in 1895. It ended with Shackleton's *Endurance* expedition of 1914–17, which coincided with the loss of innocence on the fields of Flanders as cavalry charges were cut down by machine-gun fire.

Shackleton, Scott, Amundsen, and Mawson were seeking to conquer the three "poles"—the North and South Poles and the Northwest Passage, a near-mythical sea route above Canada. Their need to challenge themselves and find out more about the world and their place in it lies at the heart of so many spheres of human endeavor and remains as true today as it was back then. It is as much about discovering what lies within as it is about triumphing over adversity.

These heroic-era expeditions also served the dual purpose of satisfying the national interests of the countries concerned and the egos of the personalities involved. These same motivations remain—although perhaps today's expeditions are done more with corporate sponsors in mind than king and country.

Shackleton first went south on the *Discovery* expedition led by Scott in 1901–03. The two men, along with Dr. Edward Wilson, got to within 720 kilometers of the South Geographic Pole. Overcome by scurvy and without sufficient food to sustain them, the trio had to turn back. They were unable to pull their sleds any farther, a problem exacerbated by their poor ski experience and dog-handling skills. Shackleton's level of debilitation was by far the worst and his subsequent evacuation home by Scott

was a slight he found difficult to live with. It began an unbridgeable rift between the two men.

Amundsen finally conquered the third pole—the Northwest Passage—in 1903–06, but not before the British Navy threw men and resources at the task in the hope of being the first to find a way through. One of its goals was to find a faster trade route from Britain to the jewel in the imperial crown, India. Attempts came thick and fast, including John Franklin's ill-fated journey. Franklin, a former governor of Tasmania, was past his prime at age fifty-nine when the expedition began in 1845. He and all 128 of his men perished—the biggest non-wartime loss of life sustained by the Royal Navy. His ships *Erebus* and *Terror* have never been found.

In 1907–09 Shackleton organized his own attempt on the South Pole—the British Antarctic Expedition, otherwise known as the Nimrod Expedition after his ship. The goal eluded his team by only 155 kilometers (97 miles), but Shackleton's decision to abandon his quest undoubtedly saved the lives of the entire party. To continue would have meant certain death. Again, had the party been more proficient dog handlers, taken a larger dog team, and dispensed with the ponies that proved to be a liability, they might have been more successful. Isn't hindsight a wonderful thing? Ultimately Shackleton's decision to turn back exemplified the compassion and fearless decision-making that came to symbolize his ability as a leader. His "surrender" was particularly brave given it ran counter to the mood of the day, where a philosophy of *pro patria mori* would soon result in millions dying for their country in the trenches of the First World War. As Shackleton said to his wife, "I thought you'd rather a live donkey than a dead lion."

Shackleton was subsequently knighted for his achievements during the Nimrod Expedition. But the trip was also notable for another major achievement. During it, Mawson, together with Alistair Mackay and Edgeworth David, reached the South Magnetic Pole. Theirs was the longest unsupported sledging journey ever undertaken and included the first ascent of the 3,800-meter volcano Mount Erebus.

The austral summer of 1911–12 was a busy period in Antarctica. Mawson set sail from Hobart to begin his Australasian Antarctic Expedition (AAE) in December 1911, just as Amundsen and Scott were racing to be the first to reach the South Pole. Amundsen reached his goal on December 14. When Scott, who toiled across the Ross Ice Shelf, arrived at the Pole on January 17, he found that the Norwegian had narrowly beaten him to his prize. He and his four men—Edward Wilson, Lawrence Oates, Henry Bowers, and Teddy Evans—started the 1,500-kilometer journey back. Evans died about a month into the return trip, and was followed by Oates, who, realizing he was a hindrance to his companions, walked out into a blizzard uttering the now immortal line, "I'm just going outside and may be some time." The remaining three men died just twenty kilometers from the final food depot that would have saved them.

Mawson would undertake his own desperate survival journey in the austral summer of 1912–13 when the Far Eastern Sledging Journey that formed part of AAE went wrong, claiming the lives of his two companions. His survival against terrible odds secured his place in the annals of Antarctic exploration history. My re-enactment of his journey of survival in 2006—in which I used the same clothing, equipment, and starvation rations—taught me the depths of resolve he must have called upon.

With Amundsen and Scott having already reached the South Pole, Shackleton embarked on the most ambitious polar expedition of all—the Imperial Trans-Antarctic Expedition (ITAE). It was a bid to cross Antarctica from the Weddell Sea coast to the Ross Sea coast in what he described as "the one great main object of Antarctic journeyings." ITAE planned to use two ships to accomplish its goal. The *Endurance*, on which Shackleton traveled, would land at a site near Vahsel Bay adjacent to the Ronne Ice Shelf in the Weddell Sea. From here Shackleton would begin his attempt to cross the continent on a route that was very similar to my unsupported expedition to cross Antarctica in 1999–2000, which left from the northernmost tip of Berkner Island on the Ronne Ice Shelf. A second ship, Mawson's former vessel the *Aurora*, would leave from Hobart under the command of Aeneas Mackintosh and land at McMurdo Sound on the Ross Sea side. Its men would then have the job of laying a series of food caches that the crossing team would access once past the Pole.

The expedition went disastrously wrong. The *Endurance* was crushed in the ice and Shackleton was forced to undertake a desperate survival bid in one of its lifeboats, the *James Caird*.

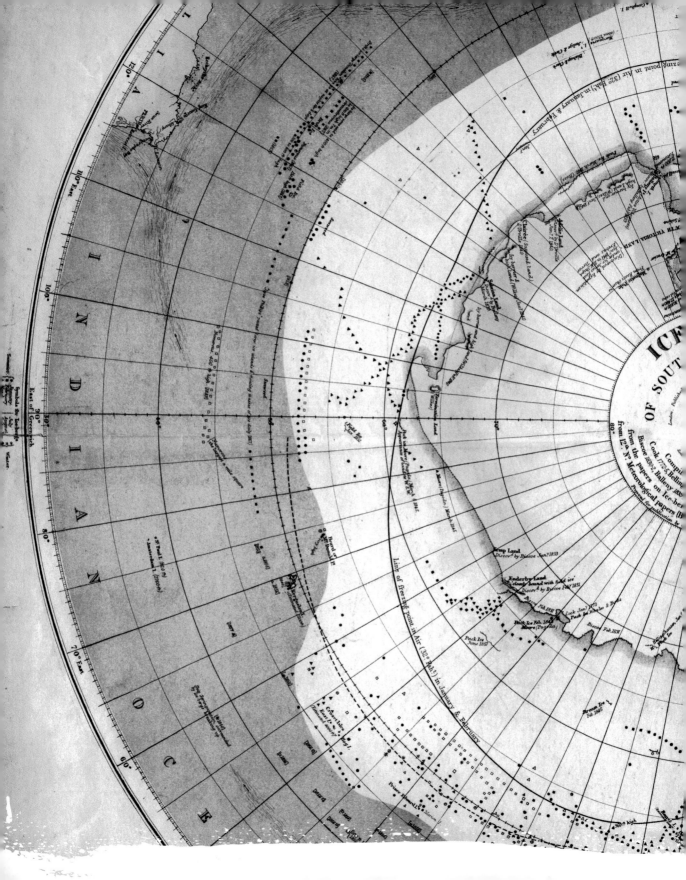

ENDURANCE

2

"Victory has 100 fathers, and defeat is an orphan."

John F. Kennedy, news conference, April 21, 1961

The plan viewed from a distance was straightforward enough: build a replica *James Caird*, take her to Antarctica on board a larger ship, hire a dedicated support vessel for the duration of the journey, select the right team, get the permits and insurance, and do it. I would finance the expedition with corporate sponsorship and sale of the film rights, supplemented by funds from fee-paying passengers who'd get a once-in-a-lifetime trip on our support vessel.

Through the help and networks of Zaz's cousin Melissa Shackleton Dann, her husband, Tom Dann, and Perry Hooks, who all lived in Washington, DC, along with the Yale World Fellows Program that saw me resident in Connecticut during the second half of 2009, I was able to get National Geographic and Discovery interested in filming the expedition. Now I could put any funds I raised toward building a boat.

After multiple trips back and forth from Yale to National Geographic's headquarters in the heart of Washington, DC, and another two all the way from Australia in early 2010, the 125-year-old company signed on. I met so many people from National Geographic—from its TV channel and production departments, its book publishing, magazine and social media arms, its speaking agency, and its expedition grants department—in an attempt to communicate the full potential of the project.

But it was all worth it. That is until a personnel change at the company coincided with a key expedition supporter and National Geographic benefactor getting cold feet. While I respected his fears for the project's safety, I had hoped National Geographic would trust my judgment. Unfortunately, his pulling out

A boat against the odds: the James Caird *on display at Shackleton's alma mater, Dulwich College.*

Previous pages: Terra Incognita.

meant National Geographic did too. To make matters worse we'd now missed our chance with Discovery. It was August 2010 and all I had to show for my lengthy efforts was a half-finished boat, a bigger mortgage, and a bruised ego.

We were back to square one, except that the tireless Seb Coulthard, my first recruit to the crew, was now working on the *Alexandra Shackleton*. All the while, the fluid nature of expedition planning meant that changes to any one set of logistics had a domino effect on all the others, keeping me second-guessing and fighting fires.

CVs were by now flooding in from people wanting to join the *Alexandra Shackleton* crew, but it was difficult to get top-notch people to commit without cast-iron guarantees that the expedition was fully funded and definitely going ahead. Without a decent broadcaster on board I couldn't get sponsors and therefore could guarantee nothing. It was a catch-22 situation: broadcasters wouldn't commit until I'd secured funding from sponsors. Also, to set an expedition date required locking in logistics providers one to two years in advance—and sponsor dollars were needed to pay their deposit fees.

Shackleton probably suffered similar problems, although he didn't have to contend with the considerable burden of bureaucracy placed on modern-day expeditions. Even with the support and understanding of the UK's Foreign and Commonwealth Office (FCO) and the South Georgian government, it was almost impossible to finalize permits until we knew the finer details of the expedition— and these would be determined to a degree by our as-yet-unknown sponsors and broadcast partner.

I edged forward on multiple fronts as best I could, financing everything myself, but it was a very lonely period of my life. I knew the risks for such projects started long before you reached the ice: risks to reputation, finances, career, and even one's marriage, as the pressures abound from throwing more and more energy and personal funds behind a project with an unknown outcome.

A turning point came in October 2010, when I joined the international engineering firm Arup on a part-time basis as a spokesman and sustainability leader. Robert Care, the chair of Asia-Pacific, and his successor, Peter Bailey, were visionaries who saw the benefits of supporting the expedition. The environmental messages of climate change and biodiversity loss that I proposed to leverage off the back of it, and the broader message of bringing to fruition something inspiring but technically and logistically challenging, paralleled what Arup was all about, making it a perfect backer for the project.

About the same time, the issue of how to get the *Alexandra Shackleton* down south was resolved. Lisa Bolton, the CEO of Aurora Expeditions, Australia's leading polar tourism operator, told me their ship, *Polar Pioneer*, took on supplies in Poland each September before heading south for the Antarctic summer season. If I could get the *Alexandra Shackleton* to Poland, it could piggyback on *Polar Pioneer* and be dropped off in Antarctica. I knew I had to make this happen even if I had to drive the trailer with the *Alexandra Shackleton* on board to Poland myself.

In the meantime, I had to convince sponsors to fund an expedition where the major cost was $300,000 for a support vessel—a legal and moral requirement in case things went wrong in the deep Southern Ocean, but not a very exciting budget line item as far as funding went. Salvation came unexpectedly in early 2012 through a contact in the nautical community of Weymouth and Portland on the southern coast of England. It was here that the *Alexandra Shackleton* was based after John Dean and Richard Reddyhoff generously allowed us to turn their state-of-the-art marina into the unofficial home of the expedition. And it was from here that Seb called me excitedly to say he'd come across a tall ship that closely resembled the *Endurance*. Maybe it could be our support vessel.

I went to meet the ship's original owner and builder at the iconic Cove House Inn, nestled behind the high shingle bank of nearby Chesil Beach, not fifty meters from "Deadman's Bay," one of the UK's most dangerous sections of coastline. Many

Shackleton planning his assault on Antarctica; some things haven't changed in a hundred years.

ships had been wrecked in the bay with great loss of life due to lee shore winds and currents driving them onshore. Just six months earlier I had no understanding of such conditions, but now I knew we would likely face a lee shore in our keel-less boat as we approached South Georgia from the southwest with winds blowing us directly onshore. With powerful surf rumbling in the background, I knew there and then that this ship, a steel-hulled barquentine that looked remarkably like the *Endurance*, was the hook needed to pull everything together.

The makers of Discovery Channel's highly successful *Gold Rush* show, Raw TV in London, loved the idea of using an *Endurance* lookalike as our support vessel. Discovery Channel Europe loved it too. Now there was an extra story angle—life aboard the *Endurance*, as well as the *Alexandra Shackleton*—although all agreed there would be no need to crush and sink our tall ship in the ice of the Weddell Sea for the sake of realism. Plus the twenty or so berths not occupied by the ship's crew and Raw's team could be made available to sponsors and other interested paying parties.

About this time, PR guru Kim McKay came on board to help with publicity and fundraising for the expedition. Not only was Kim an expert in her field, she had also worked for both National Geographic and Discovery, we had mutual friends, and she was a committed greenie who cofounded the leading environmental charity Clean Up Australia and cut her teeth doing media and PR for the BOC Challenge solo around the world yacht races—in short she was a perfect choice. It took just one serendipitous meeting in Sydney, at an event to celebrate David de Rothschild's Pacific voyage in his plastic-bottle boat *Plastiki*, and she was on board.

Expeditions are all about measuring your effort and picking your battles. It's like doubling your efforts when you know a set of tennis is there for the taking but conserving energy and conceding points cheaply when you know it is lost. With Kim on board, Arup being brilliantly supportive, Discovery Europe having committed, and Seb fitting out the *Alexandra Shackleton* in Weymouth with an army of volunteers, the stars were aligning. Suddenly we were leading two sets to one and were a service break up in the third. I decided the austral summer of 2012–13 would be our time.

Of course, I should have anticipated the match would come down to a tiebreak in the fifth set. Eternal optimism is one thing but I was, after all, trying to pull off an expedition to re-create the world's greatest survival journey during perhaps the worst recession the world has seen. Despite this, three wonderful sponsors signed on in 2012—Intrepid Travel became our naming rights sponsor; Whyte & Mackay Scotch whisky supported us with both funds and whisky (actual replica bottles of the same Mackinlay's whisky Shackleton had taken on his expedition); and St. George Bank ensured we would have enough funding to at least make the expedition happen.

With this all finalized I returned to the UK in late July 2012 to "supervise" the still formidable list of tasks needed to keep the expedition on track. I was tired but undaunted at the prospect of what lay ahead with less than six months to go: final selection of our team's sailors, sea trials, sea-survival courses and the South Georgian government's environmental and expedition briefings, final fit-out of the *Alexandra Shackleton*, answering Discovery Channel's questions, selling twenty berths aboard our support vessel, progressing the five sets of permits required for our expedition, reviewing legal aspects of contracts with sponsors and those traveling south with us, and media events in London and New York. It was a big list all right. We also had to ensure that the *Alexandra Shackleton*, on board *Polar Pioneer*, and our support vessel all left on time for their 10,000-mile journey to Antarctica. Clearly, there wasn't going to be much time for watching the London Olympics on TV.

Insurance for this whole operation, meantime, was morphing into a subject fit for a Ph.D. thesis: factors included age, level of risk exposure, and duration of that exposure on a journey that now involved two boats and thirty people re-creating the world's greatest journey of survival in the roughest ocean in the world. Shackleton would have approved of the challenge and the nine weeks in which I had to sort it all out.

Elizabeth, the boys, and I were by now house-sitting our friends Tamsin and Tom's farmhouse in England's West Country. I'm not sure what I had in mind, but I somehow thought it would be a hideaway where we could blend time as a family with my work on planning the expedition. In reality the two blended like oil and water. I'd commandeered the home office, above the old barn and away from the main house, as expedition HQ. It was a glorious spot but one where I'd already had some of my most stressful days, staring out beyond the old Tudor farmhouse to the rolling green hills of Gloucestershire. I would switch the phone off in the early hours of the morning with e-mails and messages still coming in from all over the world and reluctantly on again five hours later to see what the night had brought with it.

The *Alexandra Shackleton* needed to be finished and on board *Polar Pioneer* in Gdynia, Poland, by early September for her departure on the 20th. Our tall ship support vessel was due to start her journey south about a week later. Initially I had been charmed by the romance of using the tall ship, but for the past few months alarm bells had been ringing loudly for me, and for many serious reasons. The decision to make no changes to her schedule of traditional overseas sailing races in the immediate lead-up to her proposed departure date for Antarctica had left her way behind schedule and was indicative of how little her management appreciated the enormity of the task ahead.

To make matters worse, her skipper and his number two quit unexpectedly. There was also disagreement about fuel requirements and how to refuel safely, escalating costs, ambiguity as to how many berths were available for us to sell, and doubts over the adequacy of the clothing on board for Antarctic conditions. Having independently recruited and paid for an ice pilot and an expedition team leader, I also had to ask my good friend and polar logistics expert Howard Whelan to help the tall ship's management sort out various things I thought they should have been on top of. I couldn't help but feel they were becoming a burden I could ill afford spending time or money on. But I was committed, having invested a lot of my own money in backing their involvement. Still, I suspected that as good as they were at what they normally did, they weren't up to this challenge physically or organizationally, despite their assurances to the contrary.

Men of the sea: Dr. Robert Goodhart (left) and Philip Rose-Taylor.

I had to focus on other things, though, so I gave them a schedule of tasks that needed completion before further payments would be made and turned my attentions to getting the *Alexandra Shackleton* to Poland for her journey south.

Up in expedition HQ above the barn, I received a disturbing e-mail from *Polar Pioneer*: the frame supplied to transport the *Alexandra Shackleton* south on board *Polar Pioneer* was too big for the space set aside for her and needed to be reduced in size or she couldn't go. I swore loudly. Seb had used up some favors and $4,000 of hard-won expedition funds to get the easy-to-disassemble, color-coded frame made at the last minute to specific dimensions, and now it would need to be chopped up and adjusted when it arrived in Poland.

A few days earlier, I'd received a message asking when our expedition representatives would arrive in Poland to supervise the unloading and reloading of the *Alexandra Shackleton* onto the ship. What expedition representatives? Most of the team as it currently stood was based in Australia and working on funding, legal contracts, or selling berths aboard our support boat. Meanwhile, Howard and I were grappling with the logistics of fuel placement for the support ship, while Seb and the volunteers in Weymouth were working around the clock on finalizing fittings on the *Alexandra Shackleton*. I also had my own very long list of face-to-face meetings around the UK. Because the price tag for getting the boat to Poland on a flatbed truck was comparable to hiring a London cab to tow her there, I'd foolishly assumed the drivers could at least coordinate offloading her quayside from their truck and onto the ship without the need for us to be there. Apparently not.

We needed help, and luckily a supporter, Dr. Robert Goodhart, and Philip Rose-Taylor, a traditional sailmaker, were able to go in our stead. Two more trustworthy and capable people you'd be hard pressed to find, and, given the twinkle in Philip's eye as he left, I got the impression they loved the idea of a road trip to Europe. I just hoped these two old seadogs wouldn't be reprising some of the stuff Philip used to get up to in his youth traveling the world's oceans.

The next day we received an e-mail from *Polar Pioneer* asking for paperwork to show we had applied the biological cleaning agent Virkon to the *Alexandra Shackleton* in order to eliminate any nasties that might contaminate Antarctica. Seb immediately arranged this, making good use of his army of volunteers who were applying finishing touches to the boat at the British Navy's historic dockyard in Portsmouth following our sea trials a few weeks earlier. Philip and

Robert, meanwhile, were armed with the appropriate documentation to take to Poland, along with a letter I'd been asked to provide guaranteeing that the *Alexandra Shackleton* would be offloaded at Chile's Antarctic base, Eduardo Frei, on King George Island, although this had not yet been formally authorized. In the absence of something official from the Chilean authorities, I provided a confirmation document on expedition letterhead, knowing I had a month up my sleeve to get this signed off. At least I had the assurances of our fixer Alejo, who worked at the Frei base, that all would be well and that he would be there to take delivery of our boat. This, it turned out later, meant very little.

But we were heading in the right direction. Earlier we'd been told the boat was going to be too big and heavy for the *Polar Pioneer*. The exact dimensions had been provided to the ship's owners on a manifest from Seb indicating that the *Alexandra Shackleton* was 2.2 meters wide, not the 2.1 meters I'd told them previously. A few days prior, the Polish government had decided to load an additional shipping container, so space on the 1,000-tonne ship was now down to centimeters. I knew the slipup was mine. I hadn't realized how tight on space and weight they were on board. Luckily *Polar Pioneer*'s crane could cope with the extra weight, but it was the principle of the thing that mattered and I didn't want to test the friendship. Aurora and the crew of *Polar Pioneer* were doing us a huge favor transporting the *Alexandra Shackleton* south. Without them we'd be sunk before we got in the water.

Next Robert and Philip rang to say the captain of *Polar Pioneer* had asked to see our permits before setting sail the next morning. I broke into a cold sweat; it was another potential deal breaker. We needed five permits: a Section 3 Expedition Permit and Section 5 Ship Permit under the UK Antarctic Act, and three permits from the South Georgian authorities for ship activity, landing on South Georgia, and crossing it. All of these were still several months away. I explained to *Polar Pioneer*'s captain that we were applying for the permits and that they were all in hand but not yet finalized. The fact that he saw we were embarked on the process and speaking to the right people in accordance with the right laws put his mind at ease, but he would have been justified in refusing to let the *Alexandra Shackleton* on board. Perhaps it was the enormity of what we were trying to accomplish and how difficult it was that got us across the line.

I stared out of the barn window at the old farmhouse and could see Elizabeth and the boys in the kitchen eating without me, as had become the norm. I felt guilty that I was subjecting them to all this stress. Money had dried up and

Loading the Alexandra Shackleton *onto her mother ship,* Polar Pioneer, *for the 10,000-mile journey south from Gdynia, Poland.*

shaking the tin in the UK yielded little, so I had been eating into our mortgage for some time now, funding everything myself to the tune of tens of thousands of dollars a month. I was being as transparent as I could about it to Elizabeth while trying not to burden her unduly, but she knew me well enough to see how stressful it had become. She could also see our declining bank balance online but chose to be supportive and trust me, for which I am eternally grateful.

When we'd arrived at the farm, I'd insisted on our friend Tom telling me what jobs needed doing around the place while he was away. Reluctantly he'd mentioned a large fallen tree that needed cutting up. At one point I went into the barn and looked at the modern chainsaw that could dispatch the tree in less than a day. But in the shadows lay a rusting, heavy, blunt ax. With gladiatorial flourish I took up the ax and used it over the course of several days to batter not only the tree but also my problems into submission. Some days the pile of logs I chopped was the only tangible evidence of having made any progress; it kept me going.

I was getting tired of new problems presenting themselves each day. To (badly) paraphrase the Dalai Lama: "There are two types of problems: the ones you can overcome, in which case don't worry, and the ones you can't, in which case don't worry." I tested this philosophy to the limit most days.

I needed a break and made plans to spend three days visiting my godchildren in Brussels. It was quite something to think we were about to board a train that would take us under the English Channel while 100 meters above us, at exactly the same time, the *Alexandra Shackleton* would be on a Channel ferry. I hadn't planned it that way but that's how the dates had fallen. And I was now looking forward to having no phone reception for the half-hour tunnel journey. Just minutes before I was due to drive our vehicle onto the train, my phone rang. It was Seb. "French customs won't let the bloody boat onto the ferry and it's boarding in fifteen minutes!" "Why the hell not?" I snapped. Apparently they needed final ownership details, including my UK National Insurance number, expedition bank account, and expedition company particulars. Sensing my frustration, Seb launched into a tirade about our cross-channel neighbors, beginning with our victory at Agincourt. I cut him short, knowing I had less than eight minutes before I lost phone reception. "Wait for my call and keep the line free."

I hung up, asked Elizabeth to drive, and jumped into the passenger seat, rummaging for my laptop, which was buried under kids' toys and holiday bags. I found the bank account details and could for some unknown reason remember my UK National Insurance number even though I'd not used it for many years.

It was 10 P.M. in Australia and with three minutes until boarding time, Ramona, my PA who had been helping me out on planning issues, was my only hope. Her Canadian burr reassuringly came down the phone line, but she said it would take a couple of minutes for her to fire up her laptop and find what I needed. Elizabeth drove onto the train but mercifully it remained motionless as Ramona quoted the necessary numbers and letters to me. I called Seb and gave him the information as the train set off, hanging up seconds before we entered the tunnel. We emerged half an hour later and I turned on my phone immediately. A text message popped up saying, "All fine." It wasn't really—it was incredibly stressful—but somehow I'd got used to it.

When *Polar Pioneer* finally set sail with the *Alexandra Shackleton* on board, I was relieved beyond compare. Now I could turn my attentions to our support vessel. With just over a week to go before sailing south herself, she still needed to have an upgraded satellite communication system installed and her fire alarm system repaired, not to mention repairs to a big dent in her side obtained when a bow wave from a passing ferry caused her to break free of her moorings. Plus we still had to negotiate for 13,000 liters of diesel to be made available for her journey home from South Georgia and they had no space for the spare Zodiac I told them they needed (nor had they even purchased one). I was feeling very uneasy, but finally she left the UK bound for the Caribbean en route to Punta Arenas, Chile. Relations with her team had been strained for the past few months as we bickered over whether I was behind on payments to them or they were behind on delivering what was required in order to justify me paying them. Churchill famously said about the end of the Second World War, "Now this is not the end. It is not even the beginning of the end. But it is, perhaps, the end of the beginning." I for one felt as if I'd been in a war of attrition, and the end of the relationship was nigh.

WOODEN BOATS **3**

"They traveled in wooden boats but were iron men."

Anonymous

The onboard camera showed the gravity-defying sight of bilge water eerily snaking up the wall as the boat turned through ninety degrees, the test mannequin's neck slumping awkwardly to one side. Within seconds the same water was pooling on the ceiling and the mannequin was dramatically launched upward to join it. From our vantage point, we could see the little boat sitting improbably on her side until, ten degrees beyond vertical, she flipped suddenly onto her topside, her hull left sitting out of the water, glistening in the sun. Gunning its generator, the crane now pulled the *Alexandra Shackleton* back onto her side until her deck was not quite perpendicular to the water. In a split second she rolled back to normal-looking as a boat should.

We stood watching the capsize test, our feet firmly planted on the quayside of Portland marina in March 2012. At least we now knew the boat could go beyond vertical on its side before capsizing and didn't need to be quite vertical to roll back over from being fully inverted. In other words, she would reright more easily than she would be knocked down. That was, of course, until one realized that the waves that would help with this would find it difficult to gain purchase on the smooth, rounded hull once she was upside down. Plus the mast and sails would be vertical in the water, anchoring the boat into position. Having no keel was the issue, and capsize along with man overboard were the things we were most worried about. Basically, if the *Alexandra Shackleton* went over she would be very difficult to right. There would be no keel for a man in the water to grab on to, and even the combined weight of five men below deck would not be enough to do the job. In the meantime, any man exposed to the icy water would lose his

The reincarnation: our replica boat, the Alexandra Shackleton, at Portland.

Previous pages: On the high seas: jib and mainsail up at sunset.

ability to swim in as little as ten minutes, his muscles becoming paralyzed by the cold. Nature would have to be our savior with another big wave helping to right her, and that was down to luck.

These rerighting difficulties were because the *Alexandra Shackleton*, like the *James Caird* before her, was a whaler—a narrow, symmetrical boat with a prow and stern shaped identically, allowing it to be pulled in any direction by a harpooned whale. On such boats, the eight to ten men on board would then row the whale to shore or to a bigger whaling ship, the harpoon trace tied around the sternpost. It seemed fitting that the *James Caird* should take Shackleton's men to South Georgia, the home of Antarctic whaling, like a homing pigeon returning to the roost.

The *James Caird* was built in July 1914 by W. & J. Leslie, boatbuilders of Coldharbour Lane, near West India Docks in London. Commissioned by Frank Worsley, her skipper, and completed to his exact specifications, she was a double-ended whaler, with carvel planking of Baltic pine. This created a flush outer surface as opposed to clinker planking, where the planks overlap. Her stem and sternposts were English oak.

However, she really only became the *James Caird* we know from Shackleton's journey after the phenomenal efforts of Henry "Chippy" McNeish, carpenter on the *Endurance* and "a splendid shipwright." Helped by others among the crew after Shackleton and his men took to the ice, McNeish raised her gunwales by some thirty-five centimeters, using wood salvaged from the *Endurance*'s by-then defunct motorboat and nails from the *Endurance* herself. Shackleton knew very early on that they would have to undertake a sea voyage at some point, so he had McNeish construct whalebacks at each end and fit a pump made by photographer Frank

The same scene viewed by the team standing safely outside.

Hurley from the casing of the ship's compass. It was something that would prove invaluable in the journey ahead.

Because the *James Caird* was lightly built so as to remain "springy and buoyant" as specified by Worsley, Chippy McNeish knew he had to strengthen her spine to prevent the middle of the boat from bending up and down with the full force of the Southern Ocean, movement that could snap her in half and sink her. To do this he removed the main mast from one of the other lifeboats, the *Dudley Docker*, and bolted it to the keel of the *James Caird* to prevent her hull from "hogging and sagging" at sea. Revealing his concerns about the structural integrity of the whaler, McNeish wrote in his diary, "I am putting chafing battens on the bow of the *James Caird* to keep the young ice from cutting through as she is build of white pine which wont last long in the ice [*sic*]." All seams were caulked with lamp wick and "paid" with seal blood and artist's oil paint donated by expedition artist George Marston. This was the first recorded use of artist's paints as a form of caulk for boat seams.

At Elephant Island, Alf Cheetham and Tim McCarthy created a deck over the boat by stretching canvas over a lattice frame made by McNeish from four sledge runners and packing-case lids nailed together. Worsley recalled how, "frozen like a board and caked with ice, the canvas was sewn, in painful circumstances" by the two men whom he admiringly described as "two cheery optimists." The bow of the boat had the strongest and most watertight section of deck created by McNeish's "whaleback," which extended as far as the main mast. It was masterful work by the carpenter and regardless of his curmudgeonly nature, everyone knew what a fantastic job he had done and how indebted they were to him. Even those who found him most objectionable admitted he had worked "like a Trojan."

Incredibly, we were not the first to try to take a replica of this twenty-three-foot keel-less boat 800 nautical miles across the world's roughest ocean. We were, however, going to be the first to attempt it just as Shackleton had, with the same number of men crammed into the boat, using the same type of clothing, equipment, and traditional navigation techniques, and with no modern aids to safeguard against capsize or, in the event that it occurred, to help reright ourselves.

In 2009 Zaz had introduced me to Trevor Potts, a warm, quietly spoken, tough Geordie who, with his three crew, had attempted the "double" in 1993–94. Encountering deadly seas on the approach to Wallis Island, he was forced to sail around South Georgia to the eastern side of the island, landing at Elsie Harbor. The team attempted to cross South Georgia's mountains from Stromness to King Haakon Bay, doing the reverse of what Shackleton had done, but, due to the complexity of the terrain and a lack of food, they were forced to turn back somewhere near the Crean Glacier. Theirs was an amazing journey nonetheless and one worthy of huge respect.

Trevor was intrigued by the prospect of us attempting the journey in the exact manner Shackleton did, with six men, using only traditional navigation techniques, nonsynthetic materials for the sails and the hull of the boat, and an open cockpit to steer from using steering ropes rather than a tiller. All of these things would make our lives far more difficult, but I was adamant I wanted to do the voyage as it had been done on the *James Caird*. Add to this the fact the *Alexandra Shackleton* didn't have bunks and modern sea gear, and Trevor was impressed, although perhaps quietly skeptical as to whether the journey could really be done this way.

In early 2011, I saw a familiar, stocky figure wandering down the quayside in Ushuaia, southern Argentina. It was a mariners' crossroads if ever there was one, and, sure enough, there was Trevor. I was there to embark on a recce for our expedition, heading to South Georgia and Elephant Island as a lecturer aboard the ship *The World*, and he was on another vessel going to the Antarctic peninsula. Like ships meeting in the night—except this was a blustery afternoon in Ushuaia—we chatted enthusiastically and agreed to reconvene in the pub later on. After a few beers the normally reserved Trevor leaned over to me and said with a grim seriousness, "It's pretty scary out there, you know, Tim—noisy, cold, and rough, and big, big sea. I have to say I really don't envy you doing this but I wish you all the best and will do all I can to help." We agreed that when Trevor got back to the UK he would set down some thoughts on what he'd learned from his voyage and what he would do differently next time. As we called it a night,

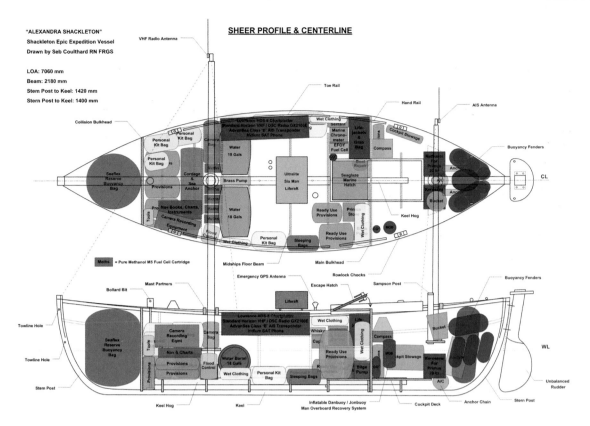

"ALEXANDRA SHACKLETON"
Shackleton Epic Expedition Vessel
Drawn by Seb Coulthard RN FRGS

LOA: 7060 mm
Beam: 2180 mm
Stem Post to Keel: 1420 mm
Stern Post to Keel: 1400 mm

SHEER PROFILE & CENTERLINE

Where do the people go? Seb's diagram of the boat's layout.

he turned to assure me, and perhaps reassure himself, that there would be no next time for him. The report I later received from Trevor proved invaluable in getting things right on board the *Alexandra Shackleton* and I am indebted to him.

Trevor mentioned that two other expeditions had attempted the double—an Irish and a German team—and sent me details. He couldn't offer introductions but suggested that decent information was publicly available on both, including, in the case of the German Arved Fuchs, his book *In Shackleton's Wake*. Former round-the-world-sailor Skip Novak, in the meantime, had supported the Irish, and both he and they were quite open to talking about the horrific experience that team had in the Southern Ocean.

The Irish team's expedition was called the South Aris and took place in January and February 1997. Their boat was named the *Tom Crean* in honor of the tough Irishman who accompanied Shackleton after begging to be part of the crew. (Originally Shackleton planned to leave Crean as a reliable right-hand man for Frank Wild on Elephant Island.) The *Tom Crean* was twenty-three feet long with a seven-foo beam, one foot wider than the *James Caird's*. She had synthetic sails, a hull of layered plywood, a tiller rather than steering ropes, and two bunks below deck, but she was certainly no pleasure craft and contained no insulation or padding.

Her tough, five-man crew of seasoned adventurers was obviously pushed to the limit on the voyage, capsizing three times in extremely bad weather (Force 10) on a course that was seemingly taking them east toward the South Orkney Islands. During each capsize, the crew found themselves upside down with the cabin half full of water. Their water-ballast transfer system allowed them to open a valve and transfer ballast weight, offsetting the center of gravity in the boat and righting it. That was all that saved them on each occasion. On getting word from their support boat, the *Pelagic Australis*, that conditions were deteriorating further, the crew abandoned the boat, with the captain the last to leave. His final act was scuttling the boat by drilling holes through the hull so she would sink and not be a danger to other ships. It was a valiant effort we were keen not to emulate.

The other attempt was that of experienced German explorer Arved Fuchs and his three shipmates in 2000. With characteristic Teutonic efficiency, his replica *James Caird* was called, well, it had to be really, the *James Caird II*. Although built from the same wood as the original, she too used synthetic sails and opted for a triangular mizzen sail rather than a squarer, gaff-rigged sail as on the original vessel. Like the other replicas, she went for a tiller rather than ropes to steer with, and her crew wore modern wet-weather and survival gear and slept in bunks. Like the *Tom Crean*, she also had an electronic rerighting system that used water ballast in the event of capsize.

Self-righting systems and tillers certainly made our steering ropes and everyone-lean-to-one-side-and-hope-for-the-best capsize-rerighting techniques seem like evolutionary dead ends by comparison, but we were determined to suffer as Shackleton had, and suffer we would.

Fuchs's team's effort was a great one, but on approaching South Georgia in stormy weather and with ice in the water, they opted to be towed into King Haakon Bay. By doing so they compromised the unsupported nature of their attempt, particularly as landing at South Georgia is one of the most difficult elements of the whole journey. We couldn't be sure the same thing wouldn't happen to us, as it would likely come down to luck with the weather. And none of us could control that.

Ironically, in among the difficulties of organizing the expedition, building a replica *James Caird* was, relatively speaking, the easy bit, particularly with the support of the James Caird Society and the existence of the original *Caird* at Dulwich College. Her life since her famous journey had not been a straightforward one. She had been brought back to England in 1919 and on

Young pretenders: the four replica Cairds to have attempted the journey since Shackleton's original.

The Sir Ernest Shackleton
Trevor Potts, 1994
Watercolor by Seb Coulthard

The Tom Crean
Paddy Barry, 1997
Watercolor by Seb Coulthard

The James Caird II
Arved Fuchs, 2000
Watercolor by Seb Coulthard

The Alexandra Shackleton
Tim Jarvis, 2013
Watercolor by Seb Coulthard

Shackleton's death in 1922 was gifted to the college by John Quiller Rowett, Shackleton's friend and the sponsor of the Quest expedition on which he died. Almost destroyed by a bomb in 1944, the *Caird* remained at the school barring a period from 1967 to 1985 when she was displayed and underwent restoration at the National Maritime Museum in Greenwich.

Zaz called me in 2008, excited at having met Nat Wilson, traditional boatbuilder extraordinaire, at the London Boat Show. She insisted I come meet him the next day and I liked him immediately. Passionate about traditional boats and with the air of someone who knew exactly what he was doing, Nat offered the free services of his college, the International Boatbuilding Training College (IBTC) in Lowestoft, Suffolk. All I needed to do was pay for the materials. Coming from one of England's best traditional boatbuilders, this was an offer I gladly accepted. Zaz asked me what the boat should be called. "I thought the *Alexandra Shackleton*," I said. "If you're comfortable with that." She certainly was.

Things progressed effortlessly and the boat was constructed on a trailer that allowed her to appear at various boat shows, where she was worked on in real time in front of fascinated onlookers. Funded by small contributions from individuals captivated by the Shackleton story, she represented tangible evidence that the expedition was on track and progressing well, although that was actually far from the truth. I went to view her a few times during 2009, on one occasion seeing two of Nat's master craftsmen grinning like children as they proudly posed for a photograph by the curved prow and sternposts as if holding the ivory tusks of a rogue elephant. "As you can see, there's great enthusiasm for the project," whispered Nat.

Seb Coulthard had joined the team in August 2010 after Calista Lucy of Dulwich College and the James Caird Society suggested he write to me to apply formally. Because funds were tight and I needed to test the mettle of any applicants, I told him about the need for people to bring either sponsorship money or personal contributions to the table. Four months and 126 letters later, Seb secured a donation of £20,000 that arrived from a corporate backer who wished to remain anonymous. Seb cleared a final hurdle when we finally met at Zaz's home on March 31, 2011. I was impressed by what I saw: the attention to detail, enthusiasm, and problem-solving ability he applied as a petty officer in the Royal Navy retrofitting Lynx helicopters for different theaters of combat could, I thought, equally be applied to retrofitting a period boat. If he could keep a helicopter in the air, by some convoluted logic I felt he could keep a replica boat afloat on the ocean. Earnest and young but articulate and passionate about Shackleton, Seb convinced me he could project-manage the *Alexandra Shackleton* into existence.

Seb immediately showed his worth. The Fleet Air Arm (the Royal Navy's air branch) had formerly operated a helicopter base at what was now John Dean and Richard Reddyhoff's Portland marina, the soon-to-be-home of the London Olympics sailing competition. Seb asked if we could use the marina to retrofit and sea-trial the *Alexandra Shackleton*. Richard immediately agreed to let us stay and use the facilities gratis for as long as required—a very generous gesture. A near miss between the *Alexandra Shackleton* and the Austrian sailing team out on the course a few months later perhaps wasn't the way to repay his confidence!

When a crowd gathered to welcome the *Alexandra Shackleton* in Portsmouth on November 7, 2011, a salty old soul, Philip Rose-Taylor, was one of the fascinated onlookers. Amazed that we were going to try to re-enact Shackleton's voyage, he introduced himself to Seb. Philip was a youthful sixty-nine-year-old, with torn-sail hair the product of a hard life spent at sea, and one of the UK's last traditional sailmakers.

Seb's fundraising efforts supplemented what I'd raised and immediately enabled Philip to produce a full set of sails and rigging for the boat. Like maritime detectives, the unlikely duo visited the *Caird* at Dulwich College. They designed a jib and mainsails and were able to model the mizzen sails on the *Caird*'s mizzen, the only sail to have survived. Every handmade stitch was perfect and used authentic materials from the period, including flax canvas, flax twine, and Manila rope. Amid the loneliness of planning all the other aspects of the expedition it was a real tonic for me to hear Seb talk about Philip's expertise in replicating the flat seams, galvanized thimbles, and Manila bolt ropes that Shackleton had on the original *Caird*. I found their enthusiasm contagious.

In January 2012, Zaz and Seb combined forces to borrow, beg, or steal equipment and boat fittings from suppliers at the London Boat Show. A great double act, they left with a haul that included safety equipment and free training from Ocean Safety, fenders from Compass Marine, and assorted classic-boat fittings such as oarlocks, brass pumps, mushroom vents, and ring bolts from Davey & Co. Seb's car became a mobile workshop, as full to the gunwales as the *Alexandra Shackleton* would surely become, and he and his navy colleagues worked long into the winter nights under fluorescent lights, amid overpowering paint, adhesive, and sealant fumes to get the boat ready.

Top: The ribs of the Alexandra Shackleton.

Bottom: Putting flesh on the bones: the larch planks.

February was given over to producing the rigging, putting the new "old" sails onto the yards and booms, and sourcing pulleys or "blocks" to control the rigging. Again the Royal Navy came to Seb and Philip's aid by welding iron rings for travelers to enable the sail to be hoisted and lowered on the mast. Authentic blocks from the period were sourced, including one made in 1839 that was reconditioned for the *Alexandra Shackleton*. Seb tested it and found it could handle a strain of more than 700 kilograms—pretty good after 174 years. Meanwhile the 1,000 kilograms of ballast we required was collected from Portland quarry in 30-kilogram bags, with the van we borrowed to transport it bowing under its strange load.

By now the *Alexandra Shackleton* had her name stenciled on her bow, so Seb and I agreed to aim for an official launch on March 18 in order to have her sea-trialed during the northern spring and ready for training during the summer. Once committed, we spread the word and locked in the date. I booked flights from Australia to come spend time undertaking capsize drills, working on the boat, and continuing to lobby for funds.

Because he had spent so much time working on the boat and knew he'd be away for months during the expedition, Seb undertook a domestic policy initiative

Maritime detective and expert traditional sailmaker Philip Rose-Taylor, hard at work.

to fit a new kitchen at his home as compensation. A simple slip while operating a circular power saw left him with a 1.5-centimeter-deep, 10-centimeter-long cut across the back of his left hand. Luckily no tendons were severed, but he required hours of surgery. In typical fashion, Seb's concern was not so much for his hand but for how he would get the boat finished by the launch date. Not wishing to burden me with it, he never told Zaz or me what had happened, instead enlisting the assistance of Paul Swain, Dean and Reddyhoff's assistant manager, who had already been a great help to the project. A fine sailor and can-do guy, Paul took on a huge workload and, with Seb's agreement, put an advertisement in the local paper requesting volunteers: engineers, boatbuilders, carpenters, and anyone else able to pick up a brush, drill, or hammer.

Finishing his kitchen with no further misadventure, Seb arrived at the marina in early March to find a group of fourteen volunteers from the local community patiently waiting to be assigned tasks. Seb thanked them profusely for coming and asked each in turn why they wanted to be involved. Without prompting they all gave a single-word answer: "Shackleton." It was a powerful moment and showed how the legend of the explorer and his achievements remained undimmed with the passage of almost a century. Among this gang of honorary shipwrights were doctors Philip Ambler and Robert Goodhart; two qualified boatbuilders, Ian Baird and Fiona Lewis; local photographer Scott Irvine; Dave and Jackie Baker; and the charming Yvonne Beven, whose house became a home away from home during the sea trials. All were tireless workers for whom no job was too big or small.

Robert Goodhart took charge of fitting the life raft and preparing the safety equipment on board. Philip Rose-Taylor set about rigging the boat and erecting the shrouds. Slowly an empty hull began to turn into a mighty craft. Fiona Lewis helped Philip get the sails on board and rigged with sheets and blocks, while Philip Ambler and Ian Baird took charge of fitting toe rails and oarlock chocks. Seb, in the meantime, wisely did anything that didn't involve power tools, given his recent track record.

Twelve days of what Seb called "expedient engineering" followed, with the team of volunteers working around the clock under the guidance of Seb, Paul, and Philip Rose-Taylor. Seb joked that the spirit of Chippy McNeish was among them during those feverish days of improvisation and working to a tight budget funded largely by my mortgage. Each time they came up with a novel solution to a problem—including, on occasion, using recycled materials—the call would go up, "We've McNeished it!" (Of course they never took shortcuts if it meant compromising safety or performance.) The final piece to be fitted on board was the hatch. Given how hard it was to open, it seemed appropriate that an outfit called Houdini Marine made it. Too fiddly for numb hands in the bitter cold of the Southern Ocean, we replaced it with something with bigger, easier-to-grasp handles.

The launch date came around quickly. Unless you looked very closely, the *Alexandra Shackleton* was essentially finished, a great testament to her hardworking team. It was amazing to see her transformed from the basic shell of a boat Nat and I had project-managed. Now she was the *James Caird* in all but name. Zaz, dignitaries, Shackleton supporters, and a large crowd of onlookers were at the launch event, with Baz Gray, our second recruit, and Seb in full military regalia. The skipper of our proposed support vessel was notably absent. It was a surprise to all of us and almost certainly a bad omen.

Neptune, god of the sea, we ask that you record our boat's name in your "Ledger of the Deep." . . . Let it be recorded, that on this day, Sunday 18th of March 2012, and forever more, this fine vessel shall be named Alexandra Shackleton. May God bless her and all who sail in her.

It was the kind of ceremony Shackleton would almost certainly have enjoyed and approved of, with Zaz laying a branch of green leaves over the deck to remind the boat that she must always return to port and Trevor Potts placing a silver coin under the mast for luck. We dispensed with breaking a champagne bottle over the prow for fear of damaging Seb's handiwork, instead splashing some of its contents on the woodwork and consuming the rest, feeling Shackleton would have approved. We toasted the four winds, but secretly I hoped for just three—easterly winds, we didn't need.

The next few months consisted of Seb continuing with sea trials and buying gear and equipment together with Paul Swain and Philip Rose-Taylor. I meanwhile returned to Australia to focus on fundraising and organize broader expedition logistics, which were now consuming all of my available time and energy. As we each focused on our niches, I left the *Alexandra Shackleton* in Seb's capable hands. He became totally immersed in old sailing technology and language, telling me on one occasion with great pride that he'd fixed a leak to the "port aft garboard strake." I was glad but had absolutely no idea what he was talking about.

Trevor Potts placing a silver coin under the mast for good luck.

Seb was a master of detail, not only sourcing period gear and equipment and establishing its provenance but taking the time to test its adequacy for the conditions. For example, our 1916 Elgin pocket watch was bought, repaired, waterproofed, and tested in a freezer to -20°C (-4° F) for twenty-four hours. Gear appeared steadily, with great stories attached to much of it, adding to the romance of the expedition. Captain Bob Turner, RN, a former captain of the ice patrol ship HMS *Endurance*, donated a Sestrel sextant similar to the Heath Hezzanith sextant used by Huberht Hudson, Shackleton's navigator on board the *Endurance*. We also acquired an E Dent compass filled with alcohol to prevent it from freezing, which was made by the same family as the original chronometer carried on the *James Caird*. And it was virtually identical to the one used by Frank Worsley.

It was an odd juxtaposition of old and new. As Seb, Paul, and Philip put in place hundred-year-old gear, Ed from Raw TV, together with marine electrician Robert Sleep, went about fitting the fixed camera rig of standard and high-definition cameras that would not have been out of place on a space shuttle.

According to Ed, it was by far the most complex camera rig ever fitted to a boat of this size (and probably the most expensive).

In June the Olympic Games sailing events began moving in to our Portland home, so the *Alexandra Shackleton* was relocated to Weymouth Marina. Best to distance her from the sleek modern Olympic boats that served only to remind us the *Alexandra Shackleton* wasn't going to win any races. More and more modern equipment now went into the boat, but none of it would give us an advantage over Shackleton—it would merely allow us to record the experience for Discovery Channel and the American PBS network. In fact, if anything, the equipment could have been seen as a disadvantage, significantly reducing the space on board while increasing electrical hazards and fire risk. Three-quarters of the carefully placed stone ballast was replaced with 560 kilograms of compact marine gel batteries, each putting out 413 amps, followed by an Automatic Identification System (AIS) transponder, chart plotter, radio, antennas, and EFOY fuel cell. God only knows what Shackleton would have made of it all.

By now Seb had commissioned a leading marine technology unit, the Wolfson Unit at Southampton University, to undertake stability calculations for the *Alexandra Shackleton*. There were two goals: we needed to know how to best position our ballast to safeguard against capsize and to understand what the chances of righting the boat would be if she did go over. The summary of their report spoke volumes:

Old and new: (left) the 360-degree infrared camera and (right) the 174-year-old wooden block.

The calculations demonstrate that the boat is not inherently self-righting. The James Caird capsized but did not remain inverted because of dynamics of the event. These calculations indicate a strong likelihood that, if the [Alexandra Shackleton] suffers a capsize and remains inverted, some action by the crew contained within the boat will enable them to right it. The scope for movement of the crew will depend on the arrangement of stowed gear and stores, and their ability to lift themselves upwards within the inverted boat. It may be desirable to conduct some trials with the boat loaded as for the proposed voyage, to quantify the level of ease or difficulty of righting the boat in this way. The consumption of water and stores will reduce the stability, and it may be desirable to counter this by filling empty containers with sea water as the voyage proceeds.

In other words, without extensive trialing we wouldn't know. What we did know was that any serious attempt to reright ourselves would require us to get ourselves as high off the ceiling as possible and hope the next wave pushed us back over. We had no sensible way of re-creating waves big enough to roll and reright the *Alexandra Shackleton* and the test sling being used to tip the boat wouldn't accurately re-create waves anyway. All of our energy would instead have to be put toward trying to ensure we didn't capsize.

In early August 2012 the team consisted of Ed Wardle, Seb, Baz, Paul Swain, and me. What were worryingly and noticeably absent to even the most casual of observers were Southern Ocean sailors. I'd lined up a series of top-notch candidates and spoken to them from Australia. Now I had to interview them in person and conduct a week of sea trials off the south coast of England to determine how well they gelled with the rest of the team. Our August sea trials therefore became as much about fine-tuning the six-man crew as about assessing the boat's seaworthiness.

With the completion of the last few jobs of fitting the compass rack in the cockpit and stowing the handmade wooden boxes containing provisions, charts, and books, we were ready to go to sea. Initial trialing began in early September in Portland's sheltered harbor, then the plan was to take to the open sea, heading east from Portland to Southampton, a distance of 100 kilometers. Seb had arranged for me and a BBC cameraman to ride in a Fleet Air Arm Lynx helicopter from Yeovilton to Portland to film her in action. As we thundered overhead I found myself looking straight down on the *Alexandra Shackleton* as we banked steeply, the wide open space beckoning through the open doors. The pilot's voice came over my headphones. "Want to go closer in?" he asked. "Yes," I replied, suggesting that we give the crew below a friendly blast of down-force from the rotors to re-create Southern Ocean conditions. He obliged with glee. Forty or so meters below us, the *Alexandra Shackleton* shook uncontrollably as she was buffeted from side to side. "You're going where in this?" asked the copilot incredulously as we moved away from the boat. "Antarctica," I replied. He and the pilot glanced at each other but said nothing.

The following evening, we were ready to begin trialing the *Alexandra Shackleton* in open water. It had taken most of the day to load her with provisions and water and go through final equipment checks. The plan was to meet and trial different potential skippers and navigators as we went east, stopping along the way to take on and drop off people. The first rendezvous point was picturesque Lulworth Cove, twenty-five kilometers to the east. Far from being an easy place to trial a boat, England's south coast has very strong tidal currents that run along it in either direction. We knew that at certain points, such as the Needles, we might even get a feel for how stable she was in big waves.

In order to meet our first potential skipper at Lulworth, we had to avoid going too far east the first evening. But a strong tide was working with us and, sure enough, even with sails trimmed and trying to tack back toward Portland as best we could, we moved eastward at several knots an hour until, in rapidly fading light, a telltale piece of flotsam passed us heading back to the west. With that, we realized the powerful tidal conveyor had turned and was beginning to push us back toward Portland. To prevent too much westerly drift all the sails now went back up, working fine initially as we trod water, the wind and current canceling each other out until the wind dropped completely about 1 A.M. Inexorably invisible forces pulled us back toward the dark, unlit band that signified Portland's rocks. It was as if we were caught in a whirlpool that wouldn't release us from its grasp until dawn, which was still four hours away. Eight miles off Portland's breakwater, the city's lights twinkling in the darkness, we realized we'd done too good a job of slowing our easterly drift—we would be on the rocks in less than three hours. It was time to take to the oars as a last resort to row east and buy some time. Baz and I leaned into them, he at the bow, me at the stern. Then, after less than five minutes of rowing, a loud crack signaled that Baz had broken his oar. Initially we laughed at the ridiculousness of the situation, but when my oar followed suit ten minutes later, we became slightly nervous. We had just one functioning oar left to keep us off the rocks (the other had a crack in the blade and a makeshift repair because we thought we'd have no need for it).

Only a long three hours of work on the oars kept us from the rocks that night as Baz and I, with some relief from the others, worked hard to balance the conflicting needs of applying power while safeguarding our last oar. At the first light of dawn, a slowing of our movement westward signaled the tide was finally going slack. Our distance from Portland's rocks was less than a mile. This wasn't even the Southern Ocean. It was Southern England and already we'd been put through our paces.

A proud patron: the Hon. Alexandra Shackleton and her namesake.

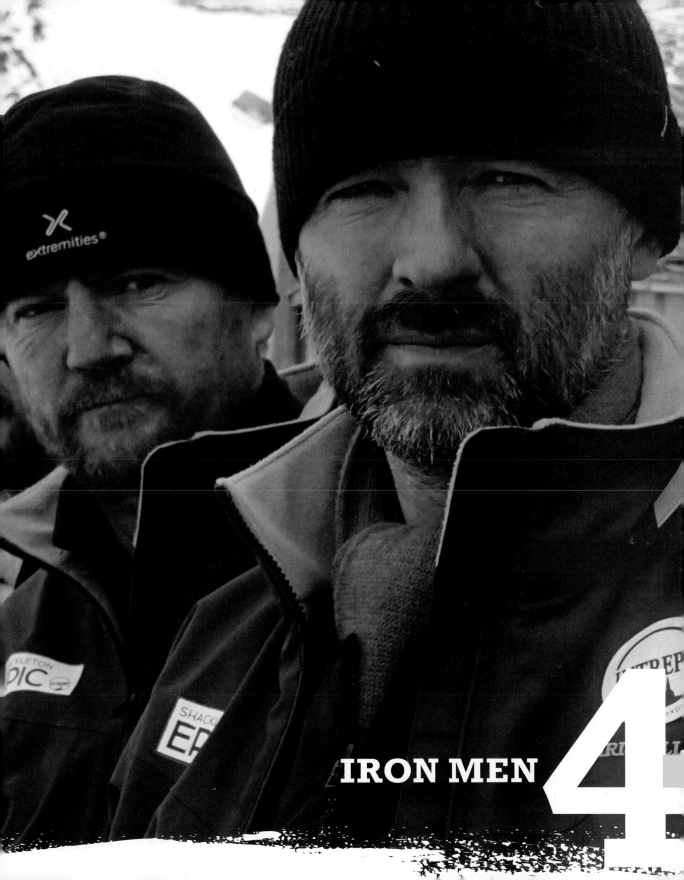

IRON MEN **4**

"Life is either a daring adventure, or nothing."

Helen Keller

On the face of it, we were two teams attempting the same journey ninety-seven years apart. On closer inspection, there were many differences over and above the passage of time. I needed to recruit just five people willing to undertake a dangerous journey in a small boat across the world's roughest ocean, followed by a climb across a mountainous island. Shackleton, on the other hand, sought twenty-seven men to fill a range of positions—everything from meteorologist, biologist, and physicist through to cook—for a journey of geographical and scientific discovery crossing the mighty continent by land, not sea. He had no idea when recruiting his team that he, together with five of his most able men, would be subjected to the ordeal of crossing the Southern Ocean in the *James Caird*. That he managed to pursue the goal of crossing the ocean in his small boat with the same rigor and determination as the original unachievable mission is just part of the Shackleton legend.

Shackleton received some 5,000 applications for the twenty-seven available places on his expedition. One of the most intriguing read:

Three strong healthy girls, and also gay and bright, and willing to undergo any hardships, that you yourself undergo. If our feminine garb is inconvenient, we should just love to don masculine attire. We have been reading all books and articles that have been written on dangerous expeditions by brave men to the Polar regions, and we do not see why men should have the glory, and women none, especially when there are women just as brave and capable as there are men.

Testing our mettle: South Georgia's challenges lay ahead of us.

Previous pages: The team.

No women went on Shackleton's expedition. A desire for historical authenticity on the part of the Shackleton family meant I could take none either.

In organizing expedition logistics, I wore an impressive carbon furrow between Australia, the UK, and the US, raising funds and sounding out suitable candidates. These included leading outdoorsmen Martin Hartley and Paul Rose as well as British sailor Pete Goss and Australia's Don McIntyre. My goal was a team split fifty-fifty between sailors and climbers. Additional skill sets—an ability to repair the boat or to film the experience—were also required. Applicants ranged from an eighty-year-old sailor, a thirteen-year-old girl, librarians, musicians, polar historians, and surgeons through to high-caliber sailors, climbers, and photographers. Fortunately, while expedition logistics and funding remained opaque, the perfect candidates revealed themselves to me at just the right time, normally in response to a specific need. Ed Wardle had come forward when many other cameramen's names were being bandied about but none seemed an obvious choice. I met him at a London cafe in June 2012 and he arrived dramatically on a powerful motorbike dressed in leathers. Quite apart from the fact he was Raw TV's choice, I could see immediately he was a solid guy, and I liked his efficient, can-do attitude and his polite, direct style. I could also see myself getting along with this Scot in the confines of a small boat and felt I could trust him. And because he had summited Everest twice, spent fifty days on his own in Alaska living off the land, and was a former UK free-diving champion (free divers hold their breath and go as deep as they can) he was an ideal candidate. The fact he had compromised personal hygiene standards and eaten absolutely anything while in Alaska was enough on its own to qualify him for life aboard the *Alexandra Shackleton*. And while I hoped we wouldn't have to call upon his free-diving skills, at least if we sank someone might survive to tell the story. As soon as I let Ed know he was on the team, he repaid me by getting to work on the camera and power systems for the *Alexandra Shackleton*, figuring out how best to film at sea and while crossing South Georgia.

The other blindingly obvious choice for the team had come a month or two earlier via a combination of reputation and an introduction from Seb, via armed forces channels. Just as Shackleton had wanted and managed to include a Royal Marine in his expedition team (in the form of Thomas Orde-Lees), we wanted to include Baz Gray, a Royal Marine. His role as mountain leader chief instructor, however, meant he trained all other mountain leaders across the UK's armed forces, had climbed everywhere, was a cold-weather expert for the Royal Navy's Antarctic patrol ship HMS *Endurance*, and had already done the crossing of South

Ed on Everest: expert cameraman and expert climber.

Georgia in modern gear and awful weather. The fact that actor Hugh Jackman and Prince Andrew had trusted him with their lives during recent charity rappels (and both had survived) also instilled a certain sense of confidence. Baz had the driest sense of humor you could imagine and a burning desire to cross South Georgia the old way—something of a rite of passage for Royal Marines. I welcomed him onto the team and he immediately took responsibility for planning all aspects of the South Georgia crossing. The only downside was Baz's predisposition to break into the opening lines of whatever song suited any given situation. But Shackleton would have loved it, I'm sure, as it seemed to tally with his sometimes unorthodox recruitment criteria. Reginald James said of his interview with Shackleton: "All that I can clearly remember of it is that I was asked if I had good teeth, if I suffered from varicose veins, and if I could sing." The interview was over within five minutes and James was appointed the expedition's magnetician and physicist.

Regardless of the fact that I required just five men to fill the *Alexandra Shackleton* and had found three of the best for their roles already, I had two main issues facing me. First, as with Shackleton, a huge team was working behind the scenes on all aspects of fundraising, logistics, equipment selection, and media management, and second, I didn't have a skipper yet. As far as the broader team went, there were people at Intrepid Travel trying to sell berths on board our support vessel, Kim McKay's Momentum2 team working on fundraising and media, and Raw TV's four-person film team and their support crew, not to mention the professional staff I'd had to find to supplement our support vessel's crew. My struggles to assemble this broader team contrasted with the ease of recruiting Ed, Baz, and Seb, three completely committed men who immediately took ownership of aspects of the expedition, speaking volumes about their commitment in a world where people are not always recognizable from the list of attributes on their CVs.

In terms of skippers, there were several candidates and two front-runners: Nick Bubb, whose name was put forward by Pete Goss, one of the world's great sailors; and Chris Stanmore-Major, who was a fantastic round-the-world team and solo sailor. Deciding which one to appoint would not be easy. It wasn't just about skill level. It was also about an ability to fit in with the rest of us and work with the other dedicated sailor on board, Paul Swain, who would be his second-in-charge. Paul had been involved in the fit-out of the *Alexandra Shackleton*, helping with all aspects of trialing and working on her as Dean and Reddyhoff's assistant manager at Portland. He was a great guy and highly accomplished, but he readily accepted that at only twenty-seven years of age and with no Southern Ocean experience, he wasn't ready to be skipper.

Now I had an immovable deadline fast approaching and if I didn't choose my skipper soon, he would barely see the *Alexandra Shackleton* let alone sail in her before she left for Antarctica in just over a fortnight's time. It would have been virtually impossible to get a world-class sailor to commit earlier, however, without assurances that the trip was fully funded and definitely happening.

I traveled to Nick's house in Lymington, near where we were due to undertake the first of our sea-survival training courses the following day in Southampton. A youthful face belied the depth of the man. Friendly, quietly self-assured but never overconfident, and with a certainty born of huge sailing experience beyond his thirty-three years, he made a good impression immediately. What I liked was that I wasn't being bowled over by a blind enthusiasm for what we were going to do; rather I was being met by a respectful but direct line of questioning about some of the tricky issues we faced. He quizzed me on whether we would continue if we lost a man, capsize and how to manage it, the role of our support vessel, and the chain of command on board. He also offered insightful suggestions on each point. I realized in an instant he had the same quality I'd seen in Ed, Baz, and Seb—an optimistic but pragmatic nature with ego in check, suggesting they would be good team players, qualities not always easy to find in those who have achieved so much. That and a healthy respect for the seriousness of what we were attempting, while not being put off by it. Rather, they aimed to rise to the challenge by offering practical solutions to overcome any problems faced. I looked for in them what others had seen in Shackleton. Worsley once wrote of his leader, "I have seen him turn pale, yet force himself into the post of greatest peril. That was his type of courage; he would do the job he was most afraid of."

Nick was a professional offshore sailor, a mechanical engineer, had raced around the world nonstop on a Maxicat in the Oryx Quest, and had competed in the Volvo Ocean Race and a whole series of other high-caliber races on virtually every class of boat. In short, he was as qualified as you could get. In fact many of the races were marathons that involved nursing a boat over long distances, and that required the kind of seamanship and mind-set I needed for our expedition. A reference from Pete Goss was no bad thing either. I told Nick the other candidates were Chris Stanmore-Major and Paul Swain and he seemed comfortable about being involved with either, although he favored Chris's greater experience.

We discovered a lot of things about the *Alexandra Shackleton* and our place in it during our five days trialing at sea: the steering ropes were fantastically heavy and cumbersome without the mechanical advantage of a tiller, the boat was sluggish and difficult to steer downwind, and living on board was going to be even more cramped, poorly ventilated, and sickness-inducing than the low expectations we already had. Plus the drinking water in our reconditioned whisky barrels had gone foul, three of our four oars had snapped, and sleeping on a floor formed of batteries was like sleeping on a pile of construction rubble. We certainly had some work to do down at the Portsmouth Historic Dockyard, Boathouse No. 4.

The Ocean Safety training, in the meantime, had taught me interesting facts about drowning and how salt water and sea spray inhalation can kill you long after you are plucked alive from the ocean. This is because your body directs fluid to your lungs in an attempt to thin the brine that fills them, causing you to drown in your own fluids. The only positive is that you can use the same process to keep yourself alive—in the absence of fresh drinking water, a salt-water enema allows your body to absorb only the fresh water and expel the salt. Images of Seb and the bilge pump suddenly made spooning seem pretty innocuous.

Seb had enlisted the help of a group of former Royal Navy personnel and they worked tirelessly to get the *Alexandra Shackleton* shipshape in ten days before she headed to Poland. These salty old sea dogs, along with Philip-Rose Taylor, Paul, Robert, Ed, and Seb, removed the rudder and most of the ballast, repainted and disinfected the boat, scrubbed all the seaweed off the hull, and fitted a variety of fixtures to the boat, including a new towing post and an additional mushroom vent. They also repaired cabling and fixtures to three of the six cameras that had been badly damaged during a lumpy part of our journey through the Needles. This was all capped off by Baz's pride and joy, his Primus cooker stowage box, being "McNeished" with a salvaged kitchen-cabinet door found in a Dumpster. Trevor Gray at the Royal Navy Historic Boat Shed in Portsmouth also helped us resolve our oar problems, letting us in to the secret that they needed to be soaked in seawater for three or four days and then oiled and sealed to prevent them from snapping.

Despite our flirtation with Portland's rocks, we didn't end up doing a skipper changeover at Lulworth Cove because Nick, who had by now committed to the project, couldn't get time off work. In characteristic fashion, Chris agreed to skipper for the whole five-day trial with Paul Swain as No. 2. They worked well together, with Paul's intimate knowledge of the *Alexandra Shackleton* and Chris's years of experience getting us through some tricky tidal conditions and, on occasion, sizable standing waves in the Needles passage. Baz was pretty seasick and Ed was unhappy about the way the cameras had been knocked about but, apart from that, things had gone well. In the meantime, Chris proved to be a great skipper and good company throughout. He was full of confidence, empowering others and not short of good stories—in all, he would be an asset to any team. There was, however, a big gap between his and Paul's experience, particularly given that Paul had not been to the Southern Ocean.

At this point Seb, having suddenly become aware of the reality of what lay ahead, told me his own sailing experience was a distant third to Paul's and that I should not be swayed by his bravado. I already had a gut feeling about Seb's limitations in terms of sailing, but it was good that he felt he could tell me. He also let me know he was mystified as to why I had decided to take him—he assumed it was because I regarded him as something of a "fixture of the boat." There was an element of that, but I also had a good feeling about his resourcefulness and ability to deliver when things got bad, as I knew they would.

Paul sensed I had reservations about his lack of Southern Ocean experience and took me to one side after the trials, volunteering to fall on his sword if need be. It was a measure of the man, him sensing I needed the experience of Chris and Nick as the two lead sailors, with Seb's boat knowledge and ability to fix things on the

Seb: committed and resourceful.

run as their backup. I turned his resignation down initially but knew he had a point. Chris meanwhile was happy to come on board, but only as skipper, whereas I saw Nick in that role and somehow couldn't see him and Chris as a natural fit in the same way as Chris and Paul. It was a quandary that was resolved for me when Nick said he would feel much happier about the project if Paul Larsen, aka Larso, was involved.

Larso was something of a known quantity to me as an Aussie with 100,000-plus sea miles and multiple world records to his name, and because I'd approached him previously, again at Pete Goss's recommendation. At the time, he had been committed to breaking the world speed record in his boat *Sail Rocket*, but winds in Namibia, where he was heading in his quest for the record, were no good from December to March, so the timing was now ideal. His CV said he had a "history of sailing some of the fastest boats ever to grace the oceans." I broke it to him gently that the *Alexandra Shackleton* was not one of them—the only way she would reach sixty-plus knots was if the crane that lifted her in and out of the water dropped her. At forty-two years of age and with a wealth of experience, Larso was a legend of the sea and importantly he knew and trusted

Nick and was keen to be involved. They were a unit and could work well together.

My weekend off in Brussels with my godchildren had not been relaxing. Major issues with our support vessel and the potential deal breaker of the *Alexandra Shackleton* not getting through French customs were still recent memories, but at least geographical distance had given me perspective on the final team selection. I gave Paul Swain the bad news that he would not be in the crew, feeling I owed it to him to tell him straight and immediately before he heard from anyone else. He was gutted but acted with the good grace I'd come to expect from him. I asked him to be our reserve sailor and he accepted. It reminded me of when Shackleton wanted the *Endurance* skippered by John King Davis, who had commanded *Aurora* during the Australian Antarctic Expedition. Davis refused, thinking the enterprise was "foredoomed," so the appointment went to Frank Worsley. By all accounts Worsley was something of an eccentric, having reportedly applied to the expedition after learning of it in a dream:

> One night I dreamed that Burlington Street was full of ice blocks and that I was navigating a ship along it. Next morning I awoke and hurried along to Burlington Street. A sign on a door caught my eye. It bore the words "Imperial Trans-Antarctic Expedition." I turned into the building, Shackleton was there, and after a few minutes' conversation he announced, "You're engaged."

It was now September 2012 and thunder and lightning flashed all around, two-meter waves rocked the launch violently and she flipped upside down, leaving us clinging to the ceiling, treading water in air pockets wherever we could find them. Regrouping with great difficulty inside the boat, we did a head count, agreed on a plan of action, ducked out into the open water, and swam to the nearby life raft. When we got in we were sitting in ankle-deep water, so we had to start bailing almost immediately. The violent movement of the life raft made us seasick and it was with great relief that I saw a helicopter's searchlight illuminate the awning of the raft as the winch man's feet appeared. One by one, we were hoisted to safety. That night the six-man team gathered in the sergeants' mess at the SB headquarters in Poole. We all agreed that the incredible simulation in the launch in the Royal National Lifeboat Association's (RNLI's) Poole wave tank was a clear reminder of how important it was to ensure we didn't capsize and that, if we did, abandoning our hundred-year-old life raft for a modern equivalent was something we really didn't want to be doing.

Top: The Rotor downforce has Baz, Woody, Seb, and Paul Swain holding on for dear life.

Bottom: Paul Larsen, aka "Larso," fastest sailor on the planet.

I returned to Australia in early October 2012, feeling for the first time that we might be a bit ahead of the curve and extremely happy with the caliber of, and unity among, the now-finalized *Alexandra Shackleton* team. That's not to say there hadn't been question marks hanging over some of my decisions but, to their credit, the guys loyally backed my choices. Ed had liked Chris Stanmore-Major both in terms of the confidence he exuded and the fact he was good in front of the camera. Seb and Baz had liked Paul Swain and probably queried the need for his replacement, although it was a measure of them and their military discipline that they never questioned me directly. Nick and Larso privately thought we needed an additional high-caliber sailor and that Seb wasn't at the required level, but nor, for that matter, were the rest of us. To my mind, Nick and Larso were a fantastic unit and sailors of the highest caliber and, most importantly, knew and trusted each other. No other combination could give us that, and we would need them to be at their best to get the job done. Seb's tireless work rate and lateral thinking made him a key team member even if he couldn't see it himself. He didn't have to. That was my job.

As well as the team being bolted down, the *Alexandra Shackleton* was on her way to Eduardo Frei, where our fixer, Alejo, would be there to receive her. We had established a great relationship with the regulatory agencies FCO (the UK Foreign and Commonwealth Office) and SGSSI (the South Georgian and South Sandwich Islands government), who were very supportive of what we were trying to do. We'd respectfully listened to all the advice they'd given and to the team's credit not a smirk or roll of the eyes had appeared when, at an SGSSI meeting, it was suggested that the *Alexandra Shackleton* could become a rat or mice colony if left unattended. There wasn't enough space on board for a single mouse!

Unfortunately, while trialing and team selection for the *Alexandra Shackleton* had gone well, the other vessels in my life were encountering headwinds. There was the small matter of acquiring and positioning 80,000 liters of diesel in Antarctica and 13,000 liters in South Georgia. Plus selling fifteen berths on our support vessel at $20,000 a head was proving troublesome, but at least Howard Whelan and I had a ten-day grace period to work on it before *Polar Pioneer* and the *Alexandra Shackleton* arrived in Buenos Aires's port, Mar del Plata, on October 12. I received an e-mail from Tomas Holik at Aurora Expeditions gently reminding me of the written guarantee I'd given to *Polar Pioneer*'s Captain Gorodnik that Alejo would rendezvous with the ship in Mar del Plata and travel aboard her to Frei to supervise the offloading of the *Alexandra Shackleton*.

Despite multiple conversations with Alejo reiterating the importance of all this, the October 12 rendezvous with Tomas in Buenos Aires didn't happen. We left Alejo multiple phone messages and e-mails suggesting a later rendezvous at Mar del Plata the next afternoon, giving clear directions how to get there. Early on Sunday, October 14, Australia time, I finally got through to Alejo on his mobile. "I'm not coming, Tim." I heard his words in slow motion, absolutely incredulous because we had a formal arrangement in place and had had numerous conversations about the importance of this rendezvous. I was furious but tried to remain calm. "Can I ask why you have decided not to be there?" "Too much ice in Fildes Bay so no boat can land at Frei," he replied. "Do you think I might have liked to have known this information?" I replied, my tone and choice use of Anglo-Saxon vernacular letting him know what I thought. Rapidly backtracking, he made some calls claiming that, due to his friendships with personnel at the Argentinian scientific base Carlini Station, formerly known as Jubany, only three nautical miles from Frei, we could offload the *Alexandra Shackleton* there. Despite the fact we were essentially a British expedition and therefore not exactly popular

Me feeling the strain but up for a challenge.

with Argentinian authorities, I was heartened by the news. When I e-mailed an increasingly concerned Tomas, he called me straight back. He knew Carlini well: its wharf's sides were like those of a bathtub and the absence of a crane or amphibious vehicle meant offloading the *Alexandra Shackleton* there was out of the question. Alejo would have known this, he said.

It was time to ditch Alejo and sort things out ourselves. There was one alternative Tomas thought might work: the Polish base of Arctowski in Admiralty Bay some twenty miles from Frei, where Tomas had spent time as a scientist many years before. Calls were made to the head of logistics at the base, Jaroslaw Roszczyk, and without hesitation he agreed we could land the *Alexandra Shackleton* there, loving the ambition of our expedition. It was fantastic news and I, of course, offered some kind of remuneration for the wonderful gesture. An e-mail came back almost immediately from Sylwia, the head of communications at the base: "Some fresh fruit and vegetables would be nice." I laughed out loud at the wonderful straightforwardness and generosity of the Poles' gesture, which wouldn't have been out of place in Shackleton's day. Four hours later, *Polar Pioneer* set sail for Arctowski.

Frei had been the first choice because it was the only base on King George Island with an airstrip and this could potentially be used to bring in the fuel our support vessel needed. Also, there was a tie-in with the Shackleton story: it was the Chileans who ultimately rescued Shackleton's men from Elephant Island in their ship *Yelcho* after the *Caird*'s voyage. When they expressed an interest in being involved in this expedition, we had told them there were two ways they could help. Ice made the first no longer an option as the *Alexandra Shackleton* was now bound for Arctowski. The second was helping transport the drums of diesel our support vessel would need to the base and allowing us to refuel there.

Regardless of their desire to help, an intense period of high-level diplomacy ensued—eight months and counting by this point. There had been countless letters, e-mails, and calls from Seb and me to the Chilean military attaché in London, the Chilean ambassador to Australia, and the head of Chile's southern fleet. This, added to Zaz's good contacts with the Chilean Navy through her grandfather's reputation, the fact we had two serving Royal Navy men in our team of six, the strong relationship between the UK and the Chilean armed forces, and my brother-in-law being based in Chile and able to help, meant we felt confident.

Our all-out assault by phone, e-mail, letters, and jumps through bureaucratic hoops brought an offer from the Chilean Navy to transport our diesel to Frei. However, it needed approval from the Chilean government, something that could only happen once we got sign-off on our environmental impact assessment (EIA)—in itself a challenge given that ship refueling is not allowed anywhere south of 60°S, a line that curiously touches no land but that demarcates Antarctica. Whichever way I put it to the FCO, which had the headache of assessing our application (as both the *Alexandra Shackleton* and our support vessel were UK-registered vessels), it was a tricky one. Refueling in the Antarctic was refueling in the Antarctic however you tried to dress it up, and negotiations were becoming protracted despite the FCO's keenness to oblige.

It was early November and I was giving a talk in Singapore, with Elizabeth and the boys having joined me. My contact at the FCO, Henry Burgess, casually mentioned to me that the Section 3 and 5 permits governing our expedition were still a way off being issued. I assumed he was referring to the issues we were having with fuel placement and proceeded to update him. Politely he cut me short. That wasn't the issue; it was the fact that our support vessel hadn't been granted permission to proceed by the Maritime Coastguard Agency (MCA). This news came as a bolt from the blue. Specifically, the issue seemed to relate to ballast calculations that hadn't taken into account factors such as ice adhering to the sails and other serious aspects I had never been made privy to. It was a pretty damning report from the MCA and coincided with a letter arriving from the Chilean government saying that, because the EIA was not issued, they could not sign off on supporting us. Now we had a Mexican standoff—or, more accurately, a Chilean standoff—because I'd been waiting to hear what level of support the

Chileans could give with fuel before determining what I put in the EIA. It meant the end of the line for our support vessel.

The longer I'd been planning things, the more I'd come to realize that the number of problems I had correlated with the number of people involved in the project. I was spending far more time on our fee-paying guests, sponsor-team members, and the film crew's logistics and insurance than on the core expedition goals. Within seventy-two hours of terminating the old support vessel, I had switched to a smaller specialist vessel, *Australis*. Luckily I had warmed up *Australis*'s team several weeks earlier when my gut told me things were going awry. It was a breath of fresh air to deal with an efficient, dedicated skipper like Ben Wallis, who knew Antarctica like the back of his hand. Despite the personal financial hit I'd taken in choosing to part company with the other boat, it was simply a decision that had to be made. I knew that, like Shackleton, you have to pursue your goals with all your being. But when changing circumstances make your goal unachievable, you need the strength to accept the situation, reset your goal, and pursue it with the same conviction as you had the original. Alexander Macklin, the surgeon on the *Endurance*, recalled Shackleton's pragmatism when the ship finally succumbed to the ice: "As always with him, what had happened had happened; it was in the past and he looked to the future. . . . Without emotion, melodrama or excitement [he] said, 'Ship and stores have gone, so now we'll go home.'"

"Going home" for me meant selling the concept of an expedition no longer supported by an *Endurance* lookalike. And not just to Raw TV and our broadcast partners, Discovery Channel in Europe and PBS in the US, but also to our four sponsors. In the meantime, Ben needed to know immediately if I was going to proceed with *Australis*, given the tightness of the deadline with which we were now working. On the positive side, the change to a smaller support vessel meant no need for fuel placement or selling berths, which was a great relief for all concerned. Sharing a small apartment in Singapore with two restless boys wasn't the most conducive environment in which to make some tough calls. Plus we were meant to be enjoying some well-earned R&R and celebrating Elizabeth's birthday. Fortunately, as always, she took things in her stride. I am so lucky to have her.

As far as Raw and our broadcast partners went, *Australis* was a better, more maneuverable vessel, which allowed closer access to us in the *Alexandra Shackleton*. Fewer people on board meant fewer things could compromise the film, plus it made things safer—all issues at the top of their list. There was always a risk that taking along paying guests to defray the cost of hiring a big vessel could end with a

Finally, we have a team: clockwise from top left, Baz, Larso, Nick, and Ed.

passenger becoming seriously injured or ill. And that could mean having to return to civilization, which would be disastrous for the film. Two long conference calls in the same number of days later, still slightly shell-shocked, they said yes.

Peter Bailey, Arup's chair in Australasia; Robyn Nixon at Intrepid; and Andy Fell at St. George Bank were all kindred spirits who understood the predicament and placed safety above all else. But they still wanted to ensure that the sponsor team got a wonderful Antarctic experience, some face time with the expedition crew, and some Shackleton leadership training. I assured them that all of these could be better delivered on *Australis* without the distractions of their having to help sail the larger vessel. They would get more training time and a more bespoke Antarctic experience, not to mention the need for their employees to take less time off work. Their final decision pending, Howard and I got to work planning a new itinerary for the sponsor team. *Australis* could take a total of fifteen on board—three permanent crew plus the four people from Raw TV meant there was room for eight sponsor guests. That would be five for Arup, one for St. George, one for Intrepid Travel, and one for the winner of a Virgin Media/

The military dry suits we used during the sea trialing.

Discovery Channel competition that had continued to run despite the loss of our larger support vessel. Okay so far. The problem was that six berths had to be set aside for the crew of the *Alexandra Shackleton* just in case we sank—a distinct possibility. To my mind, the logical solution was to give our sponsor guests their Antarctic experience before the expedition began. This could work if we could fly them into Frei on a commercial flight while the *Alexandra Shackleton* and Raw TV folks remained at Arctowski, having sailed there on *Australis*. We could then do the sea trialing that Nick and Larso wanted while our sponsors went on a ten-day tour of the Antarctic Peninsula. In principle the plan looked good as the main drivers for the sponsor team were meeting us, seeing the *Alexandra Shackleton*, visiting Antarctica, and doing all this as safely as possible, all of which I argued could be better achieved under the new arrangement. Three problems, however, now presented themselves. First, there was only one commercial flight operator—DAP Airlines—and seats were scarce as hens' teeth. Second, ten days touring the peninsula was understandably not seen to be of the same value as the original eight weeks shadowing the expedition. Third, we hadn't asked

Ed fine-tuning cameras for the challenge ahead.

Arctowski if it would be able to put up the ten-man *Alexandra Shackleton* and Raw teams. I had forty-eight hours to get it all approved.

Arctowski, in typical fashion, said yes immediately and honored the rate it might charge another country's visiting scientists—an extremely generous gesture. The eight sponsor guests were disappointed to be changing support vessel but were very supportive and wanted to remain involved. We salvaged the situation of the new proposition not being the same value as the original with the help of Robyn at Intrepid. She came up with a wonderful itinerary to Patagonia at cost price with the result that people were prepared to keep most of their money in the game. It was a good outcome given the circumstances.

Getting flights aboard DAP was not so easy. Two types of aircraft operated: a ninety-seater that individuals could not buy seats on and a six-seater Kingair that had to be chartered as a whole. We would need two people on the bigger plane and six on a chartered Kingair in order for all to arrive at the same time at Frei. The different sizes of the planes, however, meant they traveled at different speeds and the smaller plane needed better weather than the larger one in which to fly. The worry of getting both to arrive at and leave Frei on the same day and at roughly the same time kept me awake at night. I was replacing one set of complex logistics with another. The movement of the *Alexandra Shackleton* herself, with her lack of keel, seemed to be a metaphor for the planning of the expedition. Forward momentum seemed to be an illusion as the yaw of the boat sent us backward or, at best, sideways.

PROCEED 5

"Now this is not the end. It is not even the beginning of the end. But it is, perhaps, the end of the beginning."

Winston Churchill

I stared blankly at the ceiling, amazed at the hurdles I'd overcome to get this far and contemplating the challenges that lay ahead. At least I was about to have a few days off with the family, spending Christmas with my brother-in-law, Patrick, and sister-in-law, Gigi, in Santiago. But the challenge of getting expedition gear through Chilean and then Argentinian customs also loomed. The first issue on arrival would be collecting $12,000 of satellite broadband equipment purchased ten days earlier from the US and delivered directly to Santiago airport. It was needed to upgrade *Australis*'s onboard system to enable us to feed blogs and images to the media beast. Somehow I had to get this through both borders without having it confiscated and, because the funds cupboard was bare, without having to pay duty. It wouldn't be easy, especially since my bag's contents were exclusively new electronics, including an SLR waterproof camera, digital voice recorder, and two Iridium satellite phones, plus assorted Christmas presents for the kids.

At the beginning of December, as I battled boat logistics and bank balance sheets from Australia, Baz put the guys through a crash course in mountain survival and climbing techniques in Glencoe, Scotland. The goal was to build skill and confidence, particularly among the sailors on the team. A grossly oversimplified description of part of Baz's day job is taking 200 raw Royal Marine candidates out into the coldest places and putting them through their paces. Those who can handle it might get to join the Marines. Those who can't have no chance. I gave Baz free rein to do what he wanted, simply asking that he bring the team back alive. Crampon and ice-ax technique, rappelling skills, rope ascension, crevasse

Moored at Arctowski, our small boat is as ready as she's going to be for the journey ahead.

Previous pages: In Arctowski, the Alexandra Shackleton *is lifted onto an amphibious vehicle for her short journey to the sea.*

rescue, and knots were all taught over a five-day period, with several of the exercises conducted at night and without the aid of lights. Ed and Baz even elected to spend the night in subzero temperatures with no shelter, wearing the vintage gear. In Baz's words, "Everyone except Ed and I went to sleep in the comfort of their sleeping bags while we spent all night shivering, running round, shivering, running round, then we shivered and ran around some more. Morning came eventually. We had survived, with the clothing actually proving to be quite good. Except the boots—it was clear they would never be any good." Baz left Scotland "with a new confidence in the team and its ability to pull off the crossing of South Georgia."

Having thawed out after Glencoe, Nick, Ed, and Seb manfully took another one for the team, putting themselves through grueling cold-water immersion tests with leading authority Professor Mike Tipton in Portsmouth. Essentially the tests were to see how they'd cope in 2°C (35.6°F) water. Depending on body shape and size, they could remain in the water for just five to ten minutes. The first minute of submersion is the worst and we were advised, if we fell overboard, to avoid immediate overexertion so as to reduce the very real threat of cardiac arrest. (The heart is already working hard with cold shock and adrenaline.) Reassuringly, once you survive this initial shock and your temperature adjusts, then it's fine to work as fast as possible to save yourself, although ironically you lose body heat faster when you're moving than when keeping still. It was good advice but I challenge anyone who has fallen overboard in the Southern Ocean to calmly wait for his or her body to "acclimatize" before swimming like crazy to get back to the boat. That lost minute curled up in a ball and the risk of cardiac arrest would be the least of your problems. Interestingly, thermal-imaging cameras showed that such a restriction of blood flow to the extremities made frostbite and/or trench foot a real possibility during the expedition. How right that would turn out to be.

Refreshed after my Christmas break and armed with three new pairs of Christmas underpants, I headed back to Santiago airport. I was about to meet the team in Buenos Aires for the first time. Courtesy of a sterling effort on Gigi's part and a small fee, customs duty on the satellite system had been avoided and the equipment was now in a bonded warehouse. I had traveled incredibly light to start with, so shedding the kids' toys now meant my bag was virtually empty save for the contraband electronics and new undergarments. My heart sank as I saw the satellite broadband system for the first time. It was two large boxes and a massive satellite dish. Frantically I removed all its packaging and squeezed the equipment

Left: Baz leads the team training in Scotland. I simply asked he bring them back alive.

Right: Ed trying to acclimatize. Survival time at sea in 2°C water: five to ten minutes.

into my rucksack, while a mystified customs official in a Chicago Bulls T-shirt looked on. At least I hope he was an official—I gave him a $250 bond for a quasi-official looking piece of paper and he immediately vanished.

While I have experienced nothing but friendliness and hospitality in Argentina, the prospect of passing through the country instilled fear in some. On the advice of a security expert, Raw TV had at the last minute considered rerouting all of its gear and people through Brazil, amid concerns it would be heavily taxed or confiscated in Argentina by some corrupt, cheroot-smoking official. I suggested they "keep calm and carry on." Secretly, though, I had to admit that a largely British team complete with two serving military personnel going on an expedition to the disputed South Georgia and then the Falklands (aka Las Malvinas) wasn't ideal. Shackleton had left England on August 8, 1914. Despite war having broken out four days earlier, he'd received a telegram from the Admiralty telling him to "Proceed" so set off on what he termed his "white warfare." We didn't have the prospect of war hanging over our heads but the British government's decision the week before our arrival to rename a part of Antarctica claimed by Argentina Queen Elizabeth II Land wouldn't help our cause.

The plane touched down at Buenos Aires airport and, having dishonestly ticked "no" in all the boxes asking if I was bringing any new satellite phones, cameras, or other electronics into the country, I headed into the customs hall. I decided to aim for the longest line and the most flustered-looking customs official, who I hoped in the interest of time would wave me through. It was not to be and a senior official who looked like an archetypal baddie from a Tintin comic was summoned to look at the X-ray of my bag and the satellite dish. I had a speech prepared to explain the dish: I was a wildlife photographer and this was part of my equipment. I explained in the vaguest language I could muster that it was an umbrella reflector type thing needed for my photographs and of little value. I certainly didn't mention it was a brand spanking new $7,000 Iridium satellite dish. "It seems very heavy," offered the official, staring into my eyes to detect dishonesty. "Yes, there are other items inside the box as well," I replied as his junior enthusiastically took a scalpel to the multiple layers of plastic security wrap I'd encased it in specifically to deter closer inspection. Pretending to be interested only in not inconveniencing them, I warned the pair to be careful of all the loose items that were about to spill out should the packaging be cut any further. The official looked at me suspiciously and with a hint of irritation but, imagining loose items all over the conveyor belt and the resultant holdup of the two hundred or so passengers behind me, he waved me on.

Tomas Holik from Aurora Expeditions met me at the airport and we went for a steak dinner. It was late and Baz, Seb, Nick, and Larso had gone out on the town. They needed the release and it allowed me time to get to know Jamie Berry, Joe French, and Si Wagen from Raw TV. Just as Shackleton only met his expedition photographer Frank Hurley for the first time in Buenos Aires, this was my first meeting with any of the camera crew. They seemed a decent bunch: Si, a tall, laconic, roll-your-own-smokes Englishman with crazy gray hair, looked like he'd seen a few things and was full of enthusiasm for the project; Joe, young, strong, and fresh-faced, was similarly fascinated and excited by what we were about to do; and Jamie's youthful countenance belied his role as the show's on-location producer. I also met Alex Kumar, the Raw doctor, who seemed an eccentric but likable enough final member of the team.

The following morning, nursing hangovers of varying degrees, we appeared looking like a traveling mahjong team, each carrying as we were a small wooden case. Ours, however, each contained an item of breakable period equipment—compass, sextant, chronometer, and period cameras. Our twenty bags and Raw's staggering sixty-five took up a huge area of the terminal but unbelievably all got through customs unscathed. Doing a head count, we discovered that Seb was nowhere to be seen. Then a booted foot was spotted sticking out from under a pile of rucksacks where he'd fallen asleep and been thoughtfully covered over by Baz, who was presumably worried about him catching a chill.

Flying into Ushuaia, we swept low over the spectacular mountains that ring

The support crew, clockwise from top: Jamie Berry (producer) and Si Wagen (cameraman); Joe French, cameraman; Ben Wallis (the skipper of Australis*).*

the world's most southerly city. The two teams emerged from Ushuaia's small terminal and headed to their respective accommodations for the four or five days we would be there—the Raw guys to their hotel in town and us to our wonderful home away from home, Posada del Fin del Mundo, run by Ana, a friend introduced to me by Jonathan Shackleton two years earlier.

Larso and Nick spent their days in Ushuaia brushing up on astronavigation as the rest of us checked gear and began loading it aboard *Australis*. Introductions between the vessel's captain, Ben, his capable crew Skye Marr-Whelan and Magnus O'Grady, and their full complement of eleven guests were now out of the way with the arrival of our blogger extraordinaire, Jo Stewart. I liked the *Australis* crew—they were professional and multiskilled but relaxed, and nothing was too much trouble for them.

The teams had all arrived safely, *Australis* was a fine vessel, we had all of our permits, and we'd got through customs unscathed. However, all was not well in paradise. Changing support vessels meant that much of our gear had to be offloaded from our former vessel in Barbados and shipped south to our original launching point in Punta Arenas, Chile, from where it was then meant to travel to Puerto Williams for us to collect in *Australis*. Puerto Williams had been chosen as it was still in Chile, thus avoiding customs, but was on an island due southeast of us in Ushuaia and therefore en route to Antarctica by boat. For some reason the

Left: Skye at the helm of Australis.

Right: Magnus: a pair of safe hands.

drop-off had not happened. I gave the logistics firm a blast over the phone, but that wasn't going to get our gear to us, so I set out to find a solution.

The missing gear included our expedition food, sails, oars, a generator for charging our batteries, emergency flares, forty liters of 99.9 percent pure methanol for our battery recharging unit, and our medical kit. What one might call "essential items." Worst-case scenario: we would not be able to make a film and would then starve to death. Best-case scenario: we would be stuck on Elephant Island because without oars or sails the *Alexandra Shackleton* wasn't going anywhere. Trying to get on a flight to pick up the gear, much of it flammable, between Christmas and New Year's Eve and between Catholic countries with a tense relationship, was a challenge I didn't need. The solution needed to be quick and simple and in the end it was: Seb took off from the local flying club on January 2—as soon as the weather was good enough—in a two-seater plane that looked like it should be crop dusting, not going over the Andes.

In the meantime, the money from Raw for its half of the *Australis* hire fees hadn't come through. Ben took in his stride the news that money that was meant to be in his account weeks ago was still in the UK. But he politely let me know he would be within his rights not to leave port until he received the second half of the payment. I didn't blame him and got straight on the phone. A flurry of calls to the UK ended with assurances that the money would be sent to Ben's account immediately.

Later the same day I got a call from Jamie telling me to come to the naval hospital as soon as possible. I knew it meant trouble and also that, in typical documentary fashion, they would reveal nothing more until the cameras were in position to capture my arrival. Sure enough, I was met by Raw with cameras rolling. The news they'd been saving for me was that our doctor, Alex, was in a decompression chamber with the bends. Pretty impressive at sea level, I thought to myself as we approached the lozenge-shaped barometric chamber where, peering through a porthole, I could see a slightly sheepish figure sitting with an oxygen mask on his face. Determined not to give the cameras the reaction they craved, I asked how it had happened. "Surfaced too fast from his second scuba dive," replied Jamie. "What was he thinking even going on the first dive?" I asked. Hadn't we all signed papers to say we wouldn't engage in risky activity in the immediate lead-up to the expedition? It was the beginning of a painful realization that my six-man handpicked team wasn't all I had to worry about. In reality I had a team of fourteen and any one member of the group could ruin things for us all.

Jamie pressed me for a summary of how things were going. "The money's in London, the gear's in Chile, and our doctor's in a decompression chamber at an Argentinian military facility," I replied. "Apart from that, all is good." I turned and stormed off, heading for the Fin del Mundo that just about summed up how I felt.

Several hours later Seb called to say he had survived his journey to Punta in the Sopwith Camel. "There's good news and bad news . . ." he began. "Cut straight to it, Seb," I responded impatiently. "The methanol and flares aren't here, along with some of the food and the diesel generator. And the oars won't fit in the plane." I couldn't believe it. Sure enough, the oars definitely wouldn't fit inside the plane. Attaching them to the fuselage was out as was strapping them to the wings, which would apparently compromise aerodynamic lift—kind of important when you need to get over the Andes. With time against us, I told Seb to cut the oars in half at a diagonal so we could bolt them back together at Arctowski. Following several hours of frantic ringing around, Raw's fixer in Ushuaia, Roxana, found someone who used methanol in his biodiesel-making operation. After initial suspicion of what we wanted it for, forty liters were sold to us, but not before we had joined the approved supplier list to allow them to sell us the chemicals. *Australis* had a generator we could use and we bought new flares from a chandler's yard in Ushuaia, the exorbitant price reflecting the fact they knew we were desperate.

Casting off from Ushuaia delivered the first genuine sense of relief I had felt for eighteen months. After a delay caused by Alex, our doctor, getting a second opinion on his eyes, which had been troubling him since his scuba-diving misadventure, we were away, heading into the spectacular Beagle Channel, mountains on either side of us. Next stop Puerto Williams.

As we approached, we saw Seb standing quayside looking proud of himself: the oars had been repaired with metal sleeves fashioned from the hollow metal bumper bar of an old 4WD. Inspired! Puerto Williams didn't have a lot going for it except its spectacular location and the prow of the *Yelcho*, the vessel that had ultimately rescued Shackleton, standing in the middle of town like an unofficial gateway to Antarctica. For us it certainly was: from here it was across the Drake Passage to King George Island.

The bow vanished into a thick wall of green, cleaving it in two and splaying white foam from either side, our stomachs dropping dramatically as if on a roller coaster. The anemometer registered seventy-five knots of headwind as we crowded on the bridge watching the might of the ocean in awed silence. It was the end of a four-day Drake Passage crossing between Cape Horn and Antarctica, one of the

Top: Australis *moored in Ushuaia against a spectacular mountain backdrop.*

Bottom: Seb prepares for his eleventh-hour mercy flight over the Andes to rescue gear including the oars, which were too long.

roughest stretches of ocean in the world. In a bid to lighten the mood I uttered Blackadder's immortal words, "So some sort of hat is probably in order," his flippant response to the grizzled sea captain who tells him, "'Round the Cape, the rain beats down so hard it makes your head bleed." The few nervous laughs I got suggested most of us were contemplating what this would be like if we were aboard the *Alexandra Shackleton* approaching South Georgia. Later, as I was confined to my bunk in a semiconscious state, a form flew past me. It was Jo Stewart, our expedition blogger, who had been thrown violently out of her top bunk to the floor by a massive wave, luckily cushioned by the clothes and bedding that preceded her. The normally bustling communal areas of *Australis* were, by day two, eerily quiet as people tried to hold on to the contents of their stomachs. There were large vats of food still lukewarm on the stove for anyone game enough to try them, but the kitchen and living areas were now eerily reminiscent of the *Mary Celeste*. Visibility was only four miles and we wondered if the situation would be similar when trying to spot South Georgia through the mist from the *Alexandra Shackleton*. Actually, I imagined, it would be a good deal worse, being so much lower in the water.

Our ambitious, highly weather-dependent flights with sponsor guests and gear worked smoothly enough. Six people arrived at the base in Frei on the Kingair on the same day as *Australis*, linking up with the remaining two members of the team, Donald Ewen and Steve Lennon, who'd arrived on the ninety-seater two days earlier.

Frei and the adjoining base Bellingshausen, run by the Russians, were uninspiring places—a collection of drab, scruffy buildings, the only highlight being a beautiful little Eastern Orthodox church overlooking the bay. We greeted the sponsor team on the beach and everyone boarded *Australis* for the two-hour journey to Arctowski. We were all energized at the prospect of seeing the *Alexandra Shackleton* and beginning our adventures. For me, it had taken a tortuous five years to reach this point and had required me to tap into just about every emotion, skill, and ounce of resolve I possessed. As Tomas Holik had said, "With Antarctica you do not need an A and a B plan; you need also a C, D, and E plan!" How right he'd been.

Arctowski was a collection of brightly colored, prefabricated buildings but, unlike Frei, it was surrounded by majestic cliffs of rock, ice, and glaciers and set in the spectacular Admiralty Bay. Magnus and I climbed into the Zodiac to go ashore and establish contact with the locals. I felt nervous, perhaps due to the fact the whole expedition, as ever, relied on the success of this next link in the chain. For us this would be organizing all of our gear and picking a weather window to go to Elephant Island, still 100 nautical miles distant, all based on $600 of fruit and vegetables and a modest board-and-lodging rate. "Welcome to Arctowski!" shouted a lone figure on the beach as we waded ashore. It was Sylwia, with whom I'd negotiated the fruit-and-veg deal. "I have a lot of apples, oranges, and garlic for you," I offered. She laughed and suggested we get out of the cold wind. I entered the cozy, warm interior of an unmistakably Eastern European scene. In the mess-cum-bar area, pennants from those who had visited before us adorned the wood-paneled walls and bowls of sweets and nuts sat on the tables and ominously vodka-laden bar. Sylwia introduced me to Radek, an ecologist who also spoke good English, and Marek, the base commander and a bear of a man who made up for his lack of English with a permanent smile and generous hospitality.

Zodiacs plied the water between *Australis* and the base and we established ourselves in our overheated but extremely comfortable accommodations, enjoying the meat-laden meals provided by Peter, the cook. Never without his white chef's hat, apron, and broad smile below his mustache, he was a salt-of-the-earth character performing one of the most important roles on the base. The following day we were officially reunited with the *Alexandra Shackleton* for the benefit of the cameras, Baz, Seb, and I having had an illicit look at her the evening before, together with the sponsor team, who were amazed at how tiny she was. I had slept well knowing she had survived her journey from Europe and looked forward to

Arctowski base, the last frontier of comfort before we got started.

getting her in the water as quickly as possible. The next day, cameras rolling, the six-man crew strode *Reservoir Dogs* style toward the *Alexandra Shackleton*, Larso climbing inside immediately to get his first look at the boat as the sponsor group receded into the distance on board *Australis*. Larso reappeared ten minutes later with the reassuring words, "I reckon we might actually survive this thing." Seb and Ed started worked on the electrics, cameras, and batteries, Nick and Larso on the sails and rigging. Meanwhile Baz and I sorted all the clothing, climbing gear, and food.

The following morning we looked for a deep enough place to put her in the water, finding a spot 1.5 kilometers away, near the base's old oil storage tanks. A Bob the Builder convoy of vehicles—a tracked amphibious vehicle (AV), a crane, and a backhoe—disrupted the serene Antarctic scene as they trundled down the dirt track to the drop-off point. Seb, like an overprotective mother, had supervised the *Alexandra Shackleton* being lifted from the frame and placed onto the AV. Luckily his concerns about her planking being too weak to support the weight of her ballast when out of the water were unfounded. With a raise of

Yes, we can: the Alexandra Shackleton *hitching a ride on Arctowski's amphibious vehicle.*

the hand from Radek, the caravansary ground to a halt at the water's edge. The crane was gingerly positioned in deep tidal sand and kelp, its operator doing a fantastic job getting the boat into the water, undoubtedly helped by his lack of English as ten sets of conflicting instructions were uttered. With Nick and Larso aboard, Marek then dragged her out into deeper water with the Zodiac. It had taken all nine base personnel and the ten of us, but she was in and looked very much like the returning hero against a backdrop of turquoise glacial ice. "Could we fix it? Yes we had!"

Longer term, we needed a mooring closer to the base for ease of access and had found a 400-kilogram former section of concrete building foundation complete with a convenient loop of metal sticking out of it. This would allow it to be dropped onto the seabed with a mooring line attached. The crane lifted the block onto the AV and we trundled back down the road. I was wondering how we would get over the boulder-strewn beach to the open water when the nearest thing I'd seen to a smile appeared on the AV driver's normally stern face as he turned a sharp left off the track and sent us thundering down over the rocks. Making no effort to

avoid anything in our path, he powered the heavy vehicle through, grinding gears as black fumes poured from the exhaust funnel. I hadn't quite pictured this when I wrote the environmental impact assessment.

Having base Zodiacs at our disposal meant we could tow the *Alexandra Shackleton* back to her mooring at the end of each day if need be. That would gain us some time and also meant we could fine-tune our feel for her speed by checking it against a handheld GPS. The bay was a wonderful spot to trial the boat but was sheltered from big seas. We'd have to wait until the expedition proper to experience those. But at least we had a good communal space in the base's mess area where we could sit and discuss capsize and man-overboard risks.

We talked about the optimal position of *Australis* and figured two to three nautical miles was about the farthest she could afford to be from us in rough seas. This equated to fifteen to twenty minutes of her traveling at full speed in such conditions and meant she appeared more or less on the horizon. I recalled Arved Fuchs's book in which he quoted Jamie Young of the 1997 South Aris team. He said: "My greatest fear was that the boat would be rolled over by a huge sea. That happened. If the weather had gotten worse and the wind had struck at more than sixty-five knots the boat might have come apart at its weakest point in the cockpit area. If it gets really stormy out there your support vessel may as well be on the other side of the world. In any bad weather, our support vessel was miles away and struggling to take care of herself." So basically, regardless of the skill of Ben and his crew, in big seas we would be on our own. Even if she did manage to find us, the prospect of a fifty-tonne vessel like *Australis* alongside you in terrible conditions, even with an experienced skipper like Ben at the helm, wasn't good. At worst *Australis* could reduce the *Alexandra Shackleton* to driftwood.

Our relationship with the film crew while on the base was friendly enough, although their "Can you do that again for the camera?" methods got the *Alexandra Shackleton* crew's hackles up pretty quickly. We understood that they had a difficult job to do, but their constant desire for drama, overt reaction, and spontaneity wore a bit thin after a while, particularly with Baz and Larso, who resented the intrusion into our world and the time it wasted. I tried to calm my guys down a couple of times, reminding them that the film would be an important legacy, but in reality echoed their sentiment. The "You must be excited, let's see some emotion" requests were more likely to provoke a "Get the camera out of my face" reaction from us, but the film crew was slow to catch on.

I also had trouble with them laying claim to being the arbiters of authenticity.

Any perceived departure from what they regarded as Shackleton's way was challenged, with the same reason always cited: "We're doing this for you, to ensure no one is critical of you for not doing things authentically." I suspected, however, that their pedantry about clothing and equipment was more about making good TV. The more miserable we looked in our old gear, the happier they were. Authenticity levels, as far as I was concerned, should be left up to us. After all, we were the ones risking our lives in a replica twenty-three-foot boat with no modern clothing or navigational aids.

Case in point was discussions about the authenticity of the food we were eating. Shackleton had managed to take a lot of the provisions from the *Endurance* as she slowly succumbed, and had successfully supplemented this with seal and penguin meat and albatross and their eggs. Given that it was illegal to approach, let alone kill and consume, any of these animals, we had accepted there were aspects of food authenticity we simply couldn't replicate. We would eat nougat, nuts, chocolate, and tack biscuits and drink sugary milk—similar fare to the "nut food," biscuits, lump sugar, Trumilk, Bovril cubes, and Cerebos salts he had on the *Caird*. We would also eat meals of meat broth made from either pemmican (lard with beef seasoning) or army rations and cooked on a Primus kerosene stove—the equivalent to their "sledging rations." In short, our food consumption would be similar to Shackleton's without us getting too hung up about it. Shackleton took thirty days' worth of food for a journey that in the end took seventeen days, so quantity was never an issue.

The TV crew decided they wanted an experiment of half the crew eating traditional rations and half eating modern stuff in order to do a study of how tough the food was to survive on. As Ed, Baz, and I were to do the whole expedition traditionally, it seemed logical that we should be the ones to eat the old stuff, with Larso, Nick, and Seb eating more modern fare. I was skeptical of this approach, in part because I thought the three sailors would want to do the boat journey as authentically as possible, with all of us consuming the same rations. I also didn't like the idea of creating subgroups on board and didn't want to leave the door open for future claims that the sailors' modern food made their journey easier. In the end I vetoed Raw's idea, saying we would all eat the same basic food with the same calorific value as Shackleton's diet (about 4,750 to 5,000 calories a day) without obsessing over its exact makeup. It didn't go down too well, particularly with the doctor, but my decision was final.

Our team, meanwhile, was getting on well. Shackleton hadn't had the same

experience—while his crew was made up of the best men, it also included the malcontents Chippy McNeish and John Vincent, whom Shackleton had taken as part of his strategy of safeguarding the morale of those he left behind. This perceptive man-management had been noted by Lionel Greenstreet at Patience Camp when Shackleton "collected with him the ones he thought wouldn't mix with others," billeting them in his own tent. It was also highlighted when he allowed the openly anxious Thomas Orde-Lees to be ejected from one of the tents at Patience Camp by his fellow occupants due to snoring. This, Shackleton thought, would minimize the demoralizing effects of Orde-Lees's doom mongering on the other men.

The unfortunate Orde-Lees was, however, unwavering in his admiration of Shackleton, noting in his journal, "He is always careful to give his comrades the impression that he has absolute confidence in them, each in their own special sphere & yet he keeps a watchful eye on all. The reliance he places in one is certainly by far the best incentive one could have to do one's work conscientiously." My own leadership style was to intervene or ask to be consulted only if a serious situation was developing that might threaten the team, an individual, or the expedition as a whole. Beyond that, I left the sailing to Nick and Larso, boat maintenance to Seb, filming to Ed, and land-based safety to Baz, and I'm sure they appreciated it that way. Nick and Larso would inform me of what they were doing but didn't rely on my input, and Baz and Seb, as military men, were comfortable with this need-to-know way of doing things. There was, however, some tension between Seb, who was understandably very protective of all the work he had done on the *Alexandra Shackleton*, and Nick and Larso, who changed quite a few things such as some of the rigging at the last minute. Although their modifications served us well, no one disputed the fantastic job Seb did in building much of the boat, and I periodically made sure we all remembered his input.

In between our endless discussions about risks, each of us spent time personalizing and adjusting our gear, adding pockets and buttons to suit and applying grease to our gabardine jackets, trousers, and hoods to try to waterproof them. The waterproofing didn't work, but it passed the time and took our minds off the dangers ahead. I joked to Baz that we'd spent $800,000 and looked like garbage collectors. Worsley wrote that as he, Shackleton, and Tom Crean set off to cross South Georgia, "any one of [them] would have been turned out of an East End doss house." I dreaded to think what we might look like by the time we got to that stage of the journey.

It was three days before our rendezvous with *Australis* and Alex the doctor was

The Alexandra Shackleton *in her ice cradle.*

not at breakfast. I was told he'd hitched a ride on a Brazilian government vessel that morning so he could see the doctor at Frei to get a third opinion on his eyes, which continued to bother him. I understood his concern, but I was unhappy about his decision to go on many levels. If he didn't return we would have no doctor, and our expedition insurance and permits would be compromised. Plus he had made his own decision to go and had done so on a one-way trip to Frei that as far as the base was concerned meant he had made his problems their problems by turning up with no way of independently getting back to Arctowski. Alejo told me in no uncertain terms in a series of rants over the radio that he and the base were not happy with the doctor. In the end Alex was reassured, and, not happy with the third opinion, he flew to Punta Arenas for a fourth. At least he was out of the way and I knew he could get back on the empty Kingair that was coming to pick up the sponsor team in a couple of days' time. A silver lining was that he could at least collect some currency from Punta to pay our cash-only accommodation bill at Arctowski.

It was time to leave. Satellite weather information told us that in four days' time, on January 23, we would have light southerly winds at Elephant Island. This meant ideal conditions for landing at Point Wild, on the northern side of the island, protected as it is by the island's high mountains. We would need to leave for the twenty-four-hour, 100-nautical-mile journey with the first good weather, which was forecast for dawn on Monday the 21st. This was probably good too in that it meant we'd just miss out on a third alcohol-fueled Saturday night at the base, with Marek gleefully lining up vodka shots on the bar and Agatha, the base medic, ready to save anyone who went too far. Great as it was for team morale, it wasn't very conducive to being fit for what we were about to attempt. Arctowski had been wonderful on so many levels. The great group of people based there were warm, hospitable, and had a wonderful can-do attitude. Our expedition would remain forever indebted to them.

Modern communications, meanwhile, meant that financial and media commitments had followed me down to the base, and this combined with dealing with the constant demands of the TV crew meant I was always managing issues that distracted me from the massive challenge that lay ahead.

Australis appeared in the bay just off the base. When we remet the sponsor team they regaled us with stories of whale and penguin sightings, the majesty

of the Antarctic peninsula, and the things they had learned from the Shackleton leadership course, delivered by Arup's Steve Lennon using materials provided by Margot Morrell, author of the seminal text on Shackleton's leadership, *Shackleton's Way*. Relieved that their trip had been such a success, I returned with them to Frei, ensured they got on their flights back to Punta, one of which had brought the doctor back down, and after setting Alejo straight on a few things got on *Australis* and returned to Arctowski, able momentarily to breathe a sigh of relief.

Reprising the Zodiac shuttle from ten days prior, we reloaded our gear from the base. Within six hours we were ready to go, saying an emotional farewell to the nine Polish personnel at Arctowski who had suspended many of the base's day-to-day duties in order to help us during our stay. We presented them with a signed pennant and a bottle of Mackinlay's whisky and left, but not before giving Sylwia and Agatha a quick tour of the *Alexandra Shackleton* so they could see what she was like fully laden. They emerged with looks of disbelief on their faces, fearing, I suspect, that we would never be seen again. In fact, they said as much.

The *Alexandra Shackleton* left Arctowski under sail with Nick, Larso, and Ed on board. For the benefit of the film we conceded to a few hours' sailing out of the bay into open ocean, as it represented a good chance for Nick and Larso

The ice moves into Admiralty Bay, surrounding the Alexandra Shackleton *at anchor.*

to get in some more sea miles in slightly rougher conditions. In the interest of time, though, the *Alexandra Shackleton* would then be towed the rest of the way to Point Wild to ensure we got a half decent chance of landing. Making it all the way from Arctowski under sail had nothing to do with the Shackleton story anyway, whereas ensuring we left from Point Wild did.

I watched the little boat recede into the distance until, after only fifteen minutes, she was barely visible in the enormity of the seascape through which she traveled. Fleetingly she appeared as a dark silhouette against the sheer cliffs of glacial ice in the bay and it sent a shiver down my spine to see her in these surroundings. Just looking at her made me feel like I'd time-traveled back a hundred years such that I almost expected to see the *Dudley Docker* and *Stancomb Wills* alongside her. I recalled Worsley's description of the three boats crossing from the pack to Elephant Island: "Our dark sails showed in contrast with the white pack. We looked like a fleet of exploring or marauding Vikings."

For the *Alexandra Shackleton* progress was pretty slow. Even with Paul Larson, the world's fastest sailor, at the helm, we'd only managed five knots out of her in dead calm seas with strong winds coming from a perfect angle. The norm was nearer three. So at the end of Admiralty Bay we came up alongside her, picked up

The journey
to Elephant
Island begins
auspiciously:
the boat sets a
cracking pace
under tow.

Nick, Larso, and Ed, and put her on the long towline for the hundred-nautical-mile open-sea journey to Elephant Island. As we watched her from the stern of *Australis*, she kicked and rolled as if in protest at being shackled, her mast swinging madly from side to side like an out-of-control metronome. Baz said he could feel the seasickness rising already.

The open ocean was a rude shock after what we'd been used to from our sea trialing in the relative shelter of Admiralty Bay. The *Alexandra Shackleton* bucked and rocked violently with the two-meter gray swells that now passed beneath us. In anything much rougher than this the little boat was, in reality, untested. We'd told our families, sponsors, regulators, and insurers that we'd done extensive sea-trialing and had planned for most eventualities, but actually we had little idea of how the *Alexandra Shackleton* would perform in really big sea, least of all in what conditions she might capsize and how we'd recover from this. Our sea survival training had left us only too aware of how short survival time in 2°C water was. It was a very sobering thought. Out here, watching the *Alexandra Shackleton* on the high sea, it wasn't difficult to imagine what it had been like for the South Aris team upside down in their capsized boat in the dark, storm-tossed sea off the South Orkneys, terrified and fighting for their lives. It was fascinating watching the *Alexandra Shackleton* from the safety of *Australis*. But in twenty-four hours or so we would be relying on her to convey us all safely 1,500 kilometers across the forbidding gray immensity of the Southern Ocean. I turned and went back to my cabin. I'd seen enough. That was how it had been all through the planning of this expedition—I never lost sight of what lay in store but I measured my exposure to it, lest the prospect of what we were going to attempt ate into my resolve. Seeing how insignificant the *Alexandra Shackleton* looked in these surroundings wasn't very constructive—it was like watching a Jack Russell pick a fight with a Rottweiler, having no sense of its own size. The next time I saw the ocean I wanted it to be from the *Alexandra Shackleton*, not from a perspective that allowed me to see how hopelessly inadequate and small she looked in these conditions.

THE GREAT
GRAY SHROUD

6

Butting through scarps of moving marble
The narwhal dares us to be free;
By a high star our course is set,
Our end is Life. Put out to sea.

Louis MacNeice, *Thalassa*

As I watched from *Australis*'s bridge, the jagged outline of Elephant Island slowly revealed itself on the horizon in the early post-dawn light. Thirty kilometers to our right was the spectacular cloud-capped mountainous skyline of Clarence Island, rising more than 2,000 meters from the sea.

We moored tight under the cliffs of Cape Lookout on the southern tip of the island, gaining some protection from the same northwesterly winds Shackleton had experienced. The cliffs were steep and foreboding, with all but the steepest teeming with Chinstrap and Gentoo penguins, their cacophony and pungent urea overpowering from hundreds of meters away. Having used satellite weather forecasts to plan our landing at Point Wild, we now stopped here in the knowledge that a light southerly wind beckoned tomorrow—one that would maximize our chances of landing on Point Wild's north coast. This was crucial, as any failure to start from the same point as Shackleton would have compromised our expedition from the outset and would have been difficult to put behind us psychologically.

The following morning we passed the awe-inspiring natural amphitheater of rugged glacial ice that immediately precedes Cape Valentine, the easternmost point of Elephant Island. Aptly named "the Stadium," it had borne witness to the incredible feats of endurance that had unfolded at Cape Valentine. Conditions today were calm, with good visibility. Those experienced by Shackleton, however, led to him landing here in desperation, due in equal measure to concerns over the condition of his men, the big seas that prevailed, and fear of ending up on the rocks due to poor visibility. Even on a day like today landing would have been difficult.

"Growl and go": looking the part in our Shackleton-era gear. The image was taken with a 1912 Vesta pocket camera.

Previous pages: Alone with history: out in the Southern Ocean.

When we arrived at Point Wild after an hour of motoring aboard *Australis* from Cape Valentine, we felt as though we hadn't earned the right to land, such had been the relative comfort of our journey. Many were awestruck by the desolation of Point Wild, but for me it represented a place of great hope and the fulfillment of a dream to return here after my visit when planning the expedition two years ago. Our arrival brought with it an understandable amount of trepidation, and I imagined how Shackleton must have felt when he landed, knowing that someone would have to take one of the boats and undertake a desperate voyage to get help. Now, ninety-seven years later, we would undertake that same journey, for very different reasons but filled with the same dangers.

We towed the *Alexandra Shackleton* to the eastern side of Point Wild, looking for the best approach through the brash ice and rocks that lay in wait for us. The brash ranged from basketball- to fridge-sized pieces, with 90 percent of their volume lurking beneath the waterline. The larch planks on the *Alexandra Shackleton* were only 1.5 centimeters thick, so a solid impact would easily damage her. And we didn't have Chippy McNeish with us to carry out any serious repairs.

We picked our way gingerly through the small brash lumps and landed on the eastern side of the spit. Shackleton had given his youngest crew member, Perce Blackborow, the honor of alighting first on the island, having developed a paternal relationship with the young stowaway who had stolen aboard at Buenos

Elephant Island's majestic skyline rising out of the sea.

Aires. Our modern-day equivalents were the film crew who waited for us on shore cameras rolling, with whom no such paternal relationship had developed. I stepped carefully over the boulders at the seaward end of the spit to establish where the *Caird* had cast off. While the cameras lingered on the flatter shingle of the beach, I welcomed the opportunity to be away from them and quietly take in the awe-inspiring but forbidding atmosphere of the place that Hurley had gone on to describe as "truly a land where nature shows but her sullen moods." It was also a chance to reflect on all that I'd gone through to get here. Making my way carefully along the shoreline, I came across a terribly injured Chinstrap penguin hidden behind a rock. It had a large, bloody, semicircular hole in its midriff, probably courtesy of a leopard seal. The penguin looked up at me blankly, then slowly turned away to rest his chin on his chest, quietly resigned to his fate. In that moment I became instantly conscious of how lonely it would be to die here. For the penguin there would be no Shackleton-style rescue.

It was truly testament to Frank Wild's leadership that he managed to sustain the men in this inhospitable place. I wondered who had the less enviable task— Shackleton, who had to keep himself and five others motivated on a small boat but who at least controlled their destiny up to a point, or Wild, who, together with the twenty-one men left behind, was entirely reliant on the Boss's return for survival. With our being faced with a boat journey of unknown duration, I felt for Wild in not knowing how to measure his effort given that he had no idea of how long he would remain there. Wild faced this challenge magnificently, maintaining an indefatigable optimism throughout the four-month stay on the island. Shackleton wrote of him with palpable gratitude:

> Wild had reckoned that help would come in August, and every morning he had packed his kit, in cheerful anticipation that proved infectious, as I have no doubt it was meant to be. One of the party to whom I had said, "Well, you all were packed up ready," replied, "You see, boss, Wild never gave up hope, and whenever the sea was at all clear of ice he rolled up his sleeping-bag and said to all hands, 'Roll up your sleeping-bags, boys; the boss may come to-day.'"

A member of the support crew called me enthusiastically to say he'd found the precise spot from which Hurley photographed the *Caird* setting off. I boulder-hopped eagerly to the spot and, sure enough, found myself looking out to sea, the viewfinder of my camera framing exactly the scene Hurley had seen ninety-seven years prior. The image of the men manhandling the *Caird* into the surf was etched in my memory, although I could see the beach was somehow different from the way it had been then. I assumed the tide must be in, covering the sand and gravel of the Hurley photograph, but just as likely one hundred years of storms had eroded the beach.

Our *Caird* replica, meanwhile, was safely offshore beyond the breaking surf, hence our departure from Point Wild in the *Alexandra Shackleton* would be quite different from how Shackleton and his men had managed theirs. As we'd made the easy decision not to sink a perfectly good square-rigger into the pack ice of the Weddell Sea, we had only one replica whaler rather than the fleet of three that traveled down with Shackleton aboard *Endurance*, and so didn't have the luxury of being able to load our boat with ballast offshore, relaying rocks and water barrels out to her using a *Dudley Docker* or a *Stancomb Wills* as Shackleton had. Nor would it, of course, have been permissible: if everyone who had visited Point Wild over the ensuing years had taken rocks, there would be no beach left. So our little boat was already fully laden with ballast—rocks, marine batteries, food, water, sailing rig, and personal equipment—when it arrived at Elephant Island. It weighed in at more than 2.5 tonnes; three when we were on board.

Another obvious difference between the two expeditions was that we had only six men in our team; Shackleton had twenty-seven. Finding five people

As if it were only yesterday: me standing at the same point on Elephant Island where the James Caird set off for her incredible voyage (top).

crazy enough to want to join me in re-creating the *James Caird* journey was one thing. Convincing twenty-two more to wait on Elephant Island for us to rescue them would have been quite another. Our absence of manpower meant it would be nigh on impossible to keep the *Alexandra Shackleton* off the rocks were we to try to launch her from the shore, even with the small surf running today. Hurley's photographs clearly show the *Caird*, at this stage more than a tonne lighter than the *Alexandra Shackleton*, being held in position by most of the men. Our fully laden boat would have broken either us, itself, or both on the rocks if we'd tried to replicate the same beach launch.

Also, had we pushed the *Alexandra Shackleton* off the beach ourselves, we would have been soaked in the process, something we wanted to avoid at all costs. We knew from our experience at Arctowski how long it took to dry our clothes and simply couldn't afford to be plunging waist deep into icy water prior to our boat journey and spending the next few weeks trying to dry them out with body heat. As the *James Caird* was launched, McNeish and Vincent, who were standing on deck, were thrown into the swell by "a large roller" that almost capsized the boat. McNeish's trousers were wet and Vincent was soaked from head to toe. As Shackleton observed, "This was really bad luck, for the two men would have little chance of

Finally we are only minutes from departing, after four years of effort to reach this point.

drying their clothes after we had got under way." When offered an exchange of clothes by their colleagues, McNeish declined dry trousers while Vincent accepted crew member Walter How's dry trousers for his. Vincent's refusal to change out of his wet jacket "called forth some unfavourable comments as to the reason, and it was freely stated that he had a good deal of other people's property concealed about his person," according to Orde-Lees, the men having been forced to discard most of their personal items on the pack ice to save weight. How's trousers subsequently took two weeks to dry, which was certainly not something we wanted to re-create.

We left the beach by Zodiac to join the *Alexandra Shackleton* 200 meters offshore and then returned with her to the shallows to push off from there. This avoided the risk of damaging the boat or getting our crew needlessly wet. It was a balancing act. She was a strong vessel—probably stronger that the *Caird*—but not that strong.

We slipped the lines from *Australis* and rowed slowly toward Point Wild, Baz on the bow oar and me at the stern. With rocks visible in the icy water only a meter or so beneath the surface, we approached with caution. Now we were a mere minute or so away from starting our journey and the atmosphere was electric, a charged combination of awareness at the potential danger we were putting the boat and ourselves in and the knowledge that we would soon be under way. We rose and fell with each wave, all peering downward on the lookout for rocks, while occasionally looking over our shoulders to ensure no threatening wave crept up on us from behind. Gingerly we moved stern first into the narrow channel that the *Caird* had exited through more than ninety-seven years before, hissing surf all around and water cascading off the rocks. Trying to back paddle, I snagged my oar on the seabed. Quickly heaving it out of its rowlock, I used it as a punt pole, balancing the end of the blade precariously on a large, smooth boulder just meters from the beach. "That'll do it," I shouted as the hull grazed unseen objects beneath. With that, I braced myself, hoped I had good purchase, and pushed as hard as I could to send us seaward. "Row!" I yelled to Baz as he leaned into his work, and the *Alexandra Shackleton* slowly responded to our combined efforts. As the depth began to increase I jumped into the cockpit to face astern, my back out to sea, to begin a series of small strokes to start pulling the boat away from danger. The oar flexed wildly as I did so, putting Nick's repairs to the test. Slowly she steadied and made progress as Baz and I pulled hard, the clunk of small pieces of brash ice against our hull indicating we were moving out of the surf and into the sea. We had begun the penultimate chapter of the journey of our lives with the vast Southern Ocean and the mountains of South Georgia between us and destiny.

There were, of course, other major differences between our starting point for the expedition and Shackleton's. He had endured an unbelievably harsh and dangerous crossing from the pack ice just to reach Elephant Island, whereas we had been in relative comfort right up to our arrival. Worsley wrote of their circumstances several days into their bid to reach Elephant Island: "So far this boat escape had been a Rake's progress. We had rowed. We had sailed. Shackleton and I had taken turns at towing the smallest boat. We had been hindered by pack ice, head winds, currents and heavy swells. We had hauled up on the ice and escaped again. Now after 3 days of toil and exposure, without sleep, we were forty miles farther from Elephant Island. . . . The men like true British seamen ceased complaining and said grin and bear it. Growl and Go."

Nevertheless, we faced some disadvantages Shackleton did not. Most significantly, we had gone from a position of safety and predictability, where we were largely able to control our circumstances, to one filled with danger and unpredictability. In some respects Shackleton had at least arrived at a place of relative safety that may have given him a psychological advantage over us. Worsley summed it up when he wrote about the prospect of moving from the pack ice that had held them captive for so long: "In spite of our troubles and losing sleep the whole party was in good spirits, for at last we had exchanged inaction for action. We had been waiting and drifting at the mercy of the pack ice. There had been nothing that we could do to escape. Now there were more dangers and hardships, but we were working and struggling to save ourselves. We were full of hope and optimism."

Shackleton's camp at Point Wild, looking back at Elephant Island. A thin spit of land exposed to the sea on both sides, it was a better home than the pack ice but only just.

But now, regardless of our different reasons for trying to reach South Georgia, we were on parallel journeys almost a hundred years apart. It had been four years of effort to get here and a waiting game at Arctowski to secure conditions that would be right for a landing on Elephant Island. Now we were trying to get away as quickly as possible, both to escape the surf and to ensure that we didn't drift back toward the rocky coastline in whatever eddies and currents there might be.

The heavy oars moved reluctantly through the water, the weight of the *Alexandra Shackleton* from a standing start providing great resistance. The 300-meter-wide band of heavy brash ice around Point Wild—like a band of space debris orbiting a planet—didn't help, and the *Alexandra Shackleton* had to cleave a passage through it. The oar splints—cut from the hollow bumper bar with an angle grinder and bashed into position by Seb in Puerto Williams, then finessed by Nick in the Arctowski workshop—seemed to be holding out well. It was exhausting work but I knew we couldn't stop until we were a safe distance from the rocks and beyond the ice. "Shall we sail, lads?" Nick asked, just as beads of sweat started to form uncomfortably beneath my woolen layers. Baz and I gratefully stowed the oars, hoping we wouldn't need them again until the final approach into South Georgia.

The three sails were hoisted amid great anticipation; twenty minutes later, the boat had barely moved, our mood sagging along with the sails. It was unsurprising given we were in the wind shadow of a vast island, but it was disappointing just the same. Only the slow receding of the rocks at the end of Point Wild indicated any forward momentum. This was a far cry from Shackleton's departure, where they left at nearly three knots, having enough boat speed to be "shipping sea" or taking on water from the waves they crashed through. For us, by now beyond the breaking surf, there were no waves, just a gentle swell.

Hours after our departure, we could still see the dark cliffs to the left of Point Wild and the tall rock outcrop at the end of the spit. In the late evening light it looked like the silhouette of a church with a steeple. Although Shackleton makes no mention of it, the sight of it would have seemed to him a good portent. The fact they set off on their quest for salvation on Easter Monday might also have appealed to his sense of the auspicious. The way Worsley wrote about Cornwallis and Clarence islands gives an indication of the positive mood on the *James Caird*: "To the east Cornwallis and Clarence Island were revealed—two beautiful, serene and stately virgins with soft mauve wreaths and veils of misty clouds around their brows and shoulders." Unfortunately both would remain consigned to the imagination as the Southern Ocean had made Elephant Island the only practical choice of landfall.

Despite our slow pace, we set a bearing directly to the north, knowing we would be pushed to the east by the strong easterly currents that sweep around the Horn from the Pacific and out into the South Atlantic. We anticipated these would send us east at half a knot an hour—an incredible pace when one considered the top speed we'd managed from the *Alexandra Shackleton* was around only five knots at best and that at the moment we were perhaps doing one.

Heading north was part of our predetermined plan to get to the correct latitude for South Georgia and then turn in toward it on that line of latitude. It was a slightly less challenging way of navigating traditionally than trying to go straight toward South Georgia in a diagonal line. The other motivation for wanting to head due north was the fact we knew that massed pack ice lay to the immediate east of Elephant Island, perhaps as little as twenty kilometers away. It was the same motivation that governed Shackleton's decision to bear north at the beginning of his voyage. From our low vantage point on the *Alexandra Shackleton* we would be in among it virtually before we saw it, and that was somewhere we simply couldn't afford to be. Being stuck or damaged in the pack would have been disastrous.

Our northerly bearing was at the moment purely hypothetical anyway, as we weren't moving fast enough to do anything other than crab our way as much east as north for the first five or six hours. In the meantime, our next problem was slowly presenting itself—a spectacular tabular iceberg that had appeared on the horizon due north of us and right in our proposed line of travel. Perhaps four kilometers wide, it sat like a second horizon above the dark of the sea, brilliant white reflecting the dying rays of the sun.

Nick opted to head northwest slightly upwind of the iceberg. He knew that in the light southwesterly winds, sailing downwind of it might have been a bit faster for now but would potentially leave us in its wind shadow by dark. An iceberg the size of a small city, with towering 150-meter cliffs, it could render us powerless in its lee at night. If not grounded, the iceberg would continue to be moved by the wind and currents and literally run us over. Nick had demonstrated already the importance of having top-notch sailing expertise on board, as only he or Larso would have known to make such a call.

Now we just had to milk whatever forward momentum we could out of the *Alexandra Shackleton* in a bid to avoid the iceberg's westernmost tip. With darkness coming on fast, it was a slow-motion battle and one I could tell Nick and Larso were taking extremely seriously.

The iceberg was close by, its sheer imposing walls of turquoise and white ice now in shadow while its shimmering surface, smooth as a billiard table, was bathed in ethereal moonlight. We remained upwind of it but were so close now that we could hear the waves breaking against its sides. It was a beautiful yet unnerving experience as the *Alexandra Shackleton* tried to claw her way past its northern tip.

The journey from Elephant Island, save for the presence of this iceberg, had started well, with both our landing at and departure from Point Wild accomplished safely. Navigation was made easy by taking a back bearing to the rocky church and steeple silhouette of Shackleton's camp. Due to our slow progress, had there been any men left behind at Point Wild we would have been able to see their fires on the beach, as we were still less than ten kilometers offshore after more than six hours of travel. Despite the lack of speed, we were moving smoothly and steadily and felt relieved to be on our way. We'd had a celebratory tot of the Mackinlay's whisky and all was well with the world—for now at least.

As if to wish us well, a pod of whales, likely humpbacks or rights, reassuringly blew all around us, reminding us that this ocean, while feared by man, was simply home to these leviathans who lived beneath its surface. The mood was one of quiet contemplation, each of us amazed to be in the situation in which we now found ourselves—a position of real privilege to be experiencing that which Shackleton had all those years before, tinged with concern about what lay ahead.

We decided it was time to try to eat something, as we had neglected to do so in the excitement of our departure. A massive orange flame emerging from the cockpit signaled that Baz had got the Primus stove going, its glow making those of us on deck—Nick, Ed, and me—feel warmer than the -5°C (23°F) should have allowed us to. Despite the cold, we were too excited to go below and remained on watch to ensure we passed the iceberg without misadventure.

The positive feelings normally associated with food receded faster than Elephant Island as a gloved hand from below passed up a mug of congealed slop. Glistening with globules of fat, it was a dense, dark broth that wobbled to its own cadence as the boat gently rocked beneath it. It was "hoosh"—the staple of all heroic-era sledging journeys. I hadn't eaten this since I'd prepared my own for the Mawson trip in 2006, and it had taken since then to erase the memory of it. Frankly, the heroic era could have been so called based solely on the men's ability to eat this stuff, regardless of any other feats of physical or mental endurance. I brought the mug up to my mouth with trepidation, the first contact confirming the worst. Slimy, lukewarm lard with a pungent aroma overwhelmed the senses, the oil of the pemmican coating my lips with a slippery gloss that licking served only to spread until they were thickly coated like zealously overapplied lip balm. Swallowing it was even worse, as it slid down the throat far too easily, leaving an aftertaste that discouraged further consumption. My grimace spoke volumes

Baz cooking on the Primus; England expects that every man will do his duty.

but Larso needed further clarification on what I'd just experienced. "How is it on a scale of one to ten—dog shit ten, nice steak one?" "About 7.3." I laughed nervously. But I wasn't joking.

Suddenly we realized that seasickness, until now unfelt, was not far away. I managed to down half the contents of my mug before consigning the rest to the deep and cleansing my throat with a cup of hot, sugary milk. Nick's log perhaps sums it up best: "Hoosh was terrible. Boss not happy. Pan very tough to clean. . . ."

In the meantime, Nick outlined his plan for taking watches. Larso would lead one watch and Nick the other, with three hours on, three off, a slight variation on Shackleton and Worsley's strategy of two three-man teams on alternating four-hourly watches. The rest of crew were to be divided into two teams: Seb and Baz in one, me and Ed in the other. Each team would spend half their three-hour watch with Nick, and half with Larso, giving some continuity on deck. Ed and Seb were intentionally not paired up because if Seb needed to fix something, Ed would likely want to film it. I agreed too that keeping Baz and me, the two biggest crew members, on different watches was good both for space down below as well as occasions when physical grunt might be needed on deck.

Initially we tried two people in the cockpit and the remaining man below deck, on standby next to the hatch. There seemed no point in three people getting cold up above, plus there was no space for three anyway, unless the third man perched precariously on the deck itself—not a good idea given how slippery its surface was, now made even more treacherous by an unfortunate spillage of greasy hoosh. The constant likelihood of being soaked by waves, snow, and sleet or, worse still by far, the terrifying and very real prospect of falling overboard and the ruthless reality of survival time in the water, made being on deck a dicey undertaking in all but the calmest of conditions.

One by one, the cold forced all below save Larso, who remained at the helm, keeping one eye on the sail, the other on the iceberg. Life below deck was immediately confronting, the five of us expected to live, eat, sleep, and answer the call of nature in an area the size of a queen-sized bed. The acute discomfort of our quarters would have been all too familiar to Shackleton, who described in *South* how "there was no comfort in the boat. The bags and cases seemed to be alive in their unfailing knack of presenting their most uncomfortable angles to our rest-seeking bodies. A man might imagine for a moment that he had found a position of ease, but always discovered quickly that some unyielding point was impinging on muscle or bone."

During the night I awoke, cold and initially disoriented, to a pain in my right hip. Needing to move immediately to relieve the cold and discomfort, I realized I couldn't. In the half-light afforded by our two small skylights I could make out three bodies spooning as one across the width of the boat, their booted feet all resting on me, who, at six-foot-five (195 centimeters), lay along the length of the boat. Wriggling free like a rugby player after a tackle, I shuffled onto my back. This relieved the hip pain but then I felt the weight of the legs move immediately to my stomach, thighs, and chest. The stench of my reindeer-skin blanket over my face, the discomfort of having my arms pinned by my side by my crewmates' legs, and the misery of having my neck propped up awkwardly against a water barrel served as brutal reminders of where I was. I desperately needed to urinate but I could neither see the bottle nor move to look for it and so tried to put it out of my mind. Before I had time to think further about things, Larso told me it was my turn on watch, which, frankly, was welcome. My movement was met with muttering from Nick, Baz, and Ed, followed by writhing and repositioning. As I sat up I knocked my head hard on a cross spar and felt Seb's boots in my face. It was a miserable living space but at least we were dry. For now.

The blue-gray of dawn revealed we had passed the giant iceberg that had stood in our way and, although others were visible, none were in our direct path. Nick and I were on the early morning watch. He stood against the back of the cockpit, gloved hands gripping the wet ropes, elbows braced against the corners. I stood forward of him, the two of us pressed uncomfortably close together like a rider and pillion passenger on a tiny moped. In addition to its awkward intimacy, Nick couldn't actually see past me, he being the shortest and I the tallest member of the crew. But it was better than him standing in front of me, which gave him no leverage and prevented him from seeing the pennant over our left shoulders that told us wind direction. When Nick's arms eventually tired he tried sitting on the plank suspended midway up the back of the cockpit, pushing his knees into the back of mine, folding them and forcing me to sit in his lap in a position that felt even more compromising than the cozy standing configuration we'd adopted previously—something of an achievement. In addition to this indignity, Nick lost circulation in his feet, which may have been a precursor of what was to come for him. More fine-tuning of our helming positions would be required.

For now, at least, Nick and Larso were taking it in turns to steer the *Alexandra Shackleton* with a second man joining them in the cockpit, the steadily strengthening wind still coming awkwardly behind us, making the boat want to jibe continually.

The easy sailing conditions we had enjoyed for the first day and a half had gone, along with the icebergs, brash ice, and whales that had accompanied our first full day at sea. The seas were now building and at a rate faster than we could acclimatize to. Seasickness began to grip us all, with Larso and Ed choosing to vomit raucously over the side as the rest of us battled our rising nausea. The waves continued to grow in height and thickness, great, gray-green wedges of ocean pushed up by the wind. A debate briefly ensued as to how to measure them. It seemed a moot point: all I knew was from trough to crest they were approaching the height of our mast, which, at 4.9 meters, meant they were high enough.

The gale grew in strength and the boat became too much for one man to steer, so two men's wet, gloved hands were called upon to pull the rough rope and move the rudder. All the while we were being thrown about violently like rodeo cowboys. "Left! Now right!" Nick would shout as the *Alexandra Shackleton* sluggishly pulled around after a five- or six-second time lag that had us fearing she would be turned sideways and rolled down the face of the wave. We spent much of our time in the dark valleys between the waves with menacing peaks of water

Reindeer skins: yes, they molted and smelled bad, but they were welcome against the freezing cold.

peering down on us and obscuring our view of the sky. At least the two men in the cockpit were wedged so tightly they couldn't be thrown anywhere. We were doing one-and-a-half-hour shifts at the helm at this stage but after half an hour of my first shift on day three, I realized I could barely stand, feeling lightheaded and sick in roughly equal amounts. When I was back on duty later with Seb, he easily contributed more grunt than I could, even though I was twice his size. Trying to think of when I'd last eaten or drunk, I couldn't recall anything passing my lips in the past two and a half days.

Baz sensed we were all hungry to the point where poor judgment and irritability were starting to creep in. The inevitable frustration of having too little personal space and poor-quality sleep provided all the reason a hungry person needed to snap, so we tried to force down some nougat and tack biscuits with water. Unfortunately it gave us little energy and Baz, who was used to dealing

The seas building to a fever pitch were a reminder, if we needed one, of the smallness of our boat.

with the morale and well-being of large groups, realized we were slowly slipping into seasickness–induced apathy and listlessness. He decided to ignore any advice about not lighting the stove in such rough conditions in order to get a hot army-ration meal going. A brief disagreement between him and one or two of the guys followed, but he was undeterred and I fully supported his decision, realizing it was the lesser of two evils. Because the flames and boiling water were a significant danger in our confined quarters, he boiled up three lots of water to feed two men at a time. We all greedily consumed the hot meal—our first of the trip—feeling immediately better. Although the kerosene fumes may have contributed to feelings of nausea, feeding us a hot meal was the right thing to do and I was glad Baz had the presence of mind to act when he saw morale and energy levels were rapidly falling. It was one of many reasons he was on the team.

Meantime, we had now started to "ship" or take in water for the first time, with large waves periodically dumping their contents on us as the boat heeled over sharply with the sails filled, occasional gusts pushing us even farther onto our side. It felt strangely disconcerting to be leaning over steeply, knowing we had no keel to lock us into position and that our stability relied solely on the ballast on board. Getting the ballast right was crucial on a keel-less whaler: too heavy and the boat would make sluggish progress; too light and she would be like a leaf at the mercy of the elements. The issue had led to a rare disagreement between Shackleton and his fiercely loyal skipper. Writing about it almost seven years after the event, Worsley obviously still thought he was right but managed to remain tactful: "It [the ballast] was too much by about 5 cwt. The overweighting was the cause of the *Caird*'s slowness, stiffness and jerky motion. It kept us constantly wet all the passage, so causing much unnecessary misery. I demurred strongly to Sir Ernest but other counsels prevailed. He, knowing the danger of under-ballasting, went to the other extreme." We, like Shackleton, were happy to sacrifice speed and endure a more uncomfortable passage in the knowledge that the tonne of dedicated ballast plus our own weight and that of our gear gave us our best chance of remaining upright. Taking our boat out into the world's roughest ocean without conclusive capsize drills meant getting to our destination fast was not our priority. Our priority was just getting there.

TEMPEST 7

"That defiance of nature which is born of contact with humanity had hitherto sustained them, and they felt that, though alone on the vast expanse of waters, they were in companionship with others of their kind."

Marcus Clarke, *For the Term of His Natural Life*

We dropped our smallest sail—the mizzen—on the morning of our third day as the wind continued to strengthen. It was an easy enough task, completed from the relative safety of the cockpit. This was followed by reefing the main and then taking it down altogether to leave up only the handkerchief-sized jib that on its own still had the boat heeled over at a steep angle. The mainsail was a far more challenging proposition, which Larso decided to take on. He advanced carefully toward the mast as Nick watched intently, pointing the *Alexandra Shackleton* upwind to try to dampen movement—something of a token gesture given that big waves hit us hard from all angles. Even Larso, the epitome of surefooted confidence and attached with a safety line, approached the task with the respect it deserved, struggling to tug the main down as the boat tried to throw him overboard. The slippery, slanting surface of the *Alexandra Shackleton* didn't help. "Interesting," he muttered as he rejoined Nick in the cockpit after his three-minute ordeal. I recall a good friend of mine, Dave, using a similar expression when we climbed together. If he was leading and said the next bit of the climb was "interesting," I worried. The word meant far worse coming from a master of understatement.

The wind was blowing a gale, causing the lines to thrum and whistle noisily; the sea was lumpy and malevolent. Every fifth or sixth wave broke into the cockpit and the foam and spray off the wave's crests combined to soak us through. The freezing, ankle-deep water sloshing around in the cockpit saturated our leather boots and socks and left our feet numb with cold as each successive wave sent icy water straight through our jackets and trousers, drenching the woolen

Even the irrepressible Larso couldn't smile at this stage: water frozen into snow on the windward side of Larso's head covering.

Previous pages: "Something wicked this way comes . . ."

thermal layers beneath with ease and leaving us chilled to the bone. Clearly our labor-intensive attempts to waterproof our clothing at Arctowski had been futile.

By evening of the third day, we'd lost any illusions about the ability of our leather boots and gabardine outers to protect us against the cold and wet. The gabardine kept a certain amount of wind out but had the unique characteristic of letting all the water in and then retaining it. The *James Caird* crew had a similar experience, with both Shackleton and Worsley commenting on the inadequacy of their Burberry gabardine against the "all-pervading water." We could identify completely with their sodden misery and imagined them snorting in derision at the gushing words of Orde-Lees on the subject of the fabric: "This wonderful material has been so well tried and repeatedly used that no polar expedition would dream of going without suits made of it"; obviously the assessment of a man who was accustomed to remaining on solid ground and not on the Southern Ocean.

As well as increasing in size, the sea was getting more "confused" as the wind began to shift to the west, bringing with it waves from a different direction and an increased chance of inundation or capsize from a big sideways impact. This was, according to Nick and Larso, our biggest danger, even worse than our pushing into big seas or having "following" seas run over us from behind. Worsley also held that "confused, lumpy seas were far more dangerous for small boats than the straight-running waves of a heavy gale in open sea." We hadn't expected ferocious headwinds and malevolent seas. The winds during our voyage were usually coming from the south or west, just as they had for Shackleton and his crew, giving rise to the Boss's decision to make for South Georgia when he might have preferred the more populous South American mainland or the Falklands. Massive following seas could be managed by reducing sails as we had done and, if need be, by putting out our sea anchor—a parachute that acts as a brake on the boat, stopping it surfing down the face of waves and pitch-poling. But the mixed sea state we were currently experiencing was not good. Shackleton and his crew encountered similar conditions early in their journey, Worsley describing the "cross sea" they found themselves in on day two as "two seas from different directions running through or across one another, the result of two gales." "It found out our weak spots nicely," he said. 'The *Caird* was tumbling about with a hard, jerky motion and two or three bucketfuls of each sea came icy-cold over us." Now enduring our own mixed sea state, it was easy to see why nearly all of the *Caird* crew were seasick.

As we lay below, the swish and gurgle of water could be heard just centimeters from our heads through the thin skin of larch that protected us. The innocuous sound belied the speed and force with which the mighty Southern Ocean flowed past the boat, not with our forward momentum as much as its washing over us. Nick and Larso always seemed to get everything they could out of the *Alexandra Shackleton*, whatever the conditions. Skipper and navigator kept the sail flap to a minimum even in the increasing maelstrom, and I was impressed by their resilience. They weren't the biggest men on board, but their physical presence was so much greater than their size alone as they stood silent and attentive, uncomplaining sentinels on their watches.

The nonsailors subconsciously found themselves trying to read the sailors' faces to discern how concerned they were about the conditions we were facing, but they gave little away. It's an unspoken rule of expeditions, and one I always try to adhere to, that one must remain positive; if nothing positive can be said, it is best to say nothing at all. This attitude had echoes of Shackleton's own policy of deliberately repressing any negativity felt by himself or members of his crew, an attitude perhaps best exemplified as the men finally abandoned the doomed *Endurance*. According to Orde-Lees, "Everyone kept their heads splendidly. Sir Ernest's grand example inspired us all with a confidence in our leader. . . . For most of the time he stood on the upper deck holding on to the rigging, smoking a cigarette with a serious but somewhat unconcerned air." Nevertheless, Nick and Larso's increasingly businesslike attitude revealed all that needed to be known as they battled to keep the *Alexandra Shackleton* at right angles to the face of the waves. It was an unusual experience for Baz and me—more comfortable with extreme land-based expeditioning than small-boat sailing. And although Seb and Baz were in the Royal Navy, they spent much of their time on big ships. Even Ed, who had his yacht master's certificate, was really a climber at heart. We deferred to Nick and Larso's estimates of how much the *Alexandra Shackleton* could handle, but they, like us, were learning as they went along.

The seas were mixed as one wave direction tried to impose itself on another with a shift in the wind buffeting us from one side to another with increasing violence. The surface seemed alive with waves moving in all directions, but at least the predominant wind was pushing us northward. As the *Alexandra Shackleton* scudded in that direction now at a good clip—perhaps 3.5 knots with just the jib up—we realized it was a trade-off. Strong winds were good until they became too big and made the sea dangerous, thus forcing us to reduce the amount of sail we had up and slowing our forward motion.

We sailed due north as planned to offset the easterly drift of the currents that moved us relentlessly eastward at about half a knot an hour. It seemed strange to think of our puny efforts moving forward through the water while the whole column of water in which we traveled was itself moving en masse. After three days with no sun to fix our position, we weren't sure where we were exactly but we were determined not to be pushed too far to the east toward either the pack ice or the South Orkneys, where notoriously big seas reign. The South Aris team had ended up there and capsized three times in short succession before narrowly escaping with their lives. This would happen to us too if we aimed straight for South Georgia. We knew too that the pack ice would stop us dead if we ventured into it; plus, if we headed too far east we'd never be able to sail back upwind and make South Georgia. And so we aimed at some imaginary place in the open South Atlantic that we would never reach, in the knowledge that South Georgia's gravity would pull us around to the east. Like sailors of old, we would aim for the right latitude for South Georgia but keep to the west of it. When, or if, we were able to get a sun sight telling us we'd reached its latitude, we would turn right. It was a simple plan much like Shackleton's, but like all simple plans its execution would be anything but.

Certainly we appeared to be experiencing very similar weather to what Shackleton had. Worsley recalled that on "the third day it blew a hard W.S.W. gale with snow-squalls. Great torn cumulus and nimbus raced overhead. Heavy westerly seas rushing up on our port quarter swept constantly over the boat, pouring into the cockpit and coming through the canvas in little torrents soaking everything. After this, for the rest of the passage, the only dry articles in the boat were matches and sugar in hermetically sealed tins." In among the misery of our conditions we could take great solace from the fact he was here with us in spirit.

Almost twenty-four hours into the storm, the extreme cold and wet and the intense concentration required to keep us afloat meant fatigue was rapidly setting in and with it the need to share the sailing workload more evenly. A system of on-the-run coaching in rough sea steering followed for Baz, Seb, Ed, and me. Nick and Larso kept the hatch ajar to shout instructions from below, with waves routinely forcing bone-chilling green water into our cramped living area through even the smallest gap in the opened hatch, the closed air vents, and several of the forward deck planks. It was a difficult choice but we had decided that keeping the hatch ajar was safer than keeping it shut. It improved ventilation, reducing the chances of lethal conditions like hypoxia, caused by carbon monoxide fumes from the stove, and it also helped keep seasickness at bay, while allowing the backup man on watch to keep an eye on the helmsman to ensure nothing happened to him. It also meant when big waves crashed over the deck the helmsman could shout "Bail!" and the backup man could spring into action with the bilge pump that was situated below deck near the hatch. Arm-wrestling the bilge pump with frozen hands was exhausting, particularly in our state, and it repeatedly became blocked with food and other debris that washed around in the bilges. Seb, grim faced as he held back seasickness, took it apart and, on one occasion, retrieved a small plastic tie-wrap from the blocked outlet valve, a reminder of the twenty-first century. Seb was always taking on jobs uncomplainingly and with real application. It was one of the reasons I had chosen him.

The downside of leaving the hatch ajar was that the giant waves crashing over us continually drenched whoever was sitting near it below deck, regardless of the speed with which the helmsman tried to close it as the next wave approached. The man on the helm in the cockpit would, of course, be utterly soaked but that was his lot. The routine of the hatch dramatically slamming shut every few minutes became one of the sounds of life aboard the *Alexandra Shackleton*, always in response to the helmsman spotting something big coming our way. It was always with some relief that the hatch opened again, indicating that the helmsman had survived the latest onslaught.

Was Shackleton with us in spirit? A mistaken double exposure taken with our 1912 Vesta pocket camera shows two boats.

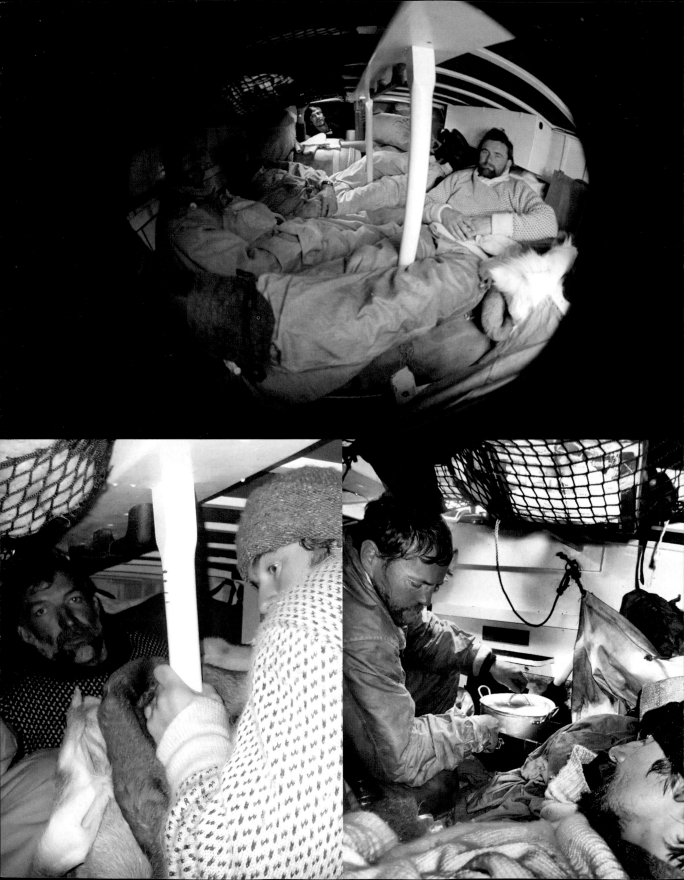

For the nonsailors among us, the learning curve that our watches at the helm presented us with was as steep as the rock faces many of us were more used to climbing. The weight of responsibility to guide the boat through this environment that raged and rolled darkly around us was exhausting. We could identify with Worsley's detailed descriptions of how he and his crew were soaked by "bucketfuls" of ice-cold water with every large wave and then had to start pumping frantically. Perhaps the most telling and symbolic observation he made was, 'This was our baptism—the beginning of the ordeal by water." His analogy rang true for sailors and nonsailors alike in this watery battlefield we found ourselves in.

Fear of being capsized was real now, especially as the boat heeled over at such a steep angle that the mast almost touched the waves. No one really spoke about how well our capsize drill would hold up and what the repercussions would be if we went over. Frankly, it didn't bear thinking about if capsize occurred with the hatch fully open. The living space below deck was our only means of keeping the boat afloat, the air contained within it providing the boat with virtually all its buoyancy save an air bag we'd inflated in the forward bulkhead. So staying afloat relied on us keeping the hatch shut in big seas. But if we did go over and the boat failed to reright within ten minutes, the only thing for it would be to open the two mushroom air vents and deliberately flood the boat with about 200 to 300 liters of water to help offset the righting moment. In other words, our weight plus that of the water we'd deliberately let in could be transferred to one side and, with a very big wave's assistance, hopefully we'd be knocked back upright. The resistance of the sails under the water would be a hindrance, of course, and all this would be going on in freezing cold, huge seas, and pitch darkness below deck, with the possible injury of one or more men and a certain degree of panic setting in. We'd also not know if the one or perhaps two men in the cockpit were still attached to the boat or even alive. We also knew that if we let in too much water through the vents, or if either vent should be stuck open, which they were frequently, we would flood completely and either drown or sink. And if rerighting took longer than fifteen minutes, the man in the water would likely have lost the ability to swim in the extreme cold and sunk beneath the waves.

This survival time in the water could, of course, be increased tenfold if we wore our neoprene survival suits. At Arctowski, we'd trialed how long it would take us all to put these on below deck and one after the other jump overboard as if abandoning ship. Although each man abandoning the "stricken vessel" left more space for those who remained, the fear would also be increasing with each minute that passed as the boat got closer to sinking. In calm conditions in Admiralty Bay, the best we could manage was twenty-nine minutes to don our suits. Unless we were in them way in advance of us needing to abandon ship or the boat capsizing or sinking, we were dead. I'm sure most of us thought about these options but said nothing. Better to focus on not going over in the first place.

As it happened, even though the seas were now large and imposing, we had decided there was no point in two men getting soaked at the helm, particularly as the limited space didn't allow for the second man to help much with steering anyway. Having trialed the second man on watch sitting low in the cockpit by the helmsman's feet, we'd discovered that this cramped position, although out of the wind, induced bad seasickness and meant sitting in a bath of icy water while being doused with water from the waves above. We simply had to increase our recovery time down below and dry our clothes off with our own body heat and that of the others, much as Shackleton had done. In Worsley's words: "Fortunately the bow space, when gained, was penetrated only by the heaviest seas, so that the sleeping bags did not get thoroughly wet for two days, and then we could generate a certain amount of warmth before the next extra heavy sea." Other than the water that periodically penetrated the forward planks and got through the closed air vents, we remained relatively dry in the forward section of the boat, although the trade-off was that it was airless and the worst spot in the boat for seasickness. It was here that Ed's camera console remained in waterproof housing and he somehow managed to operate it despite the seasickness in all but the worst conditions.

The *Alexandra Shackleton* was terrible to steer, with her steering ropes, just like those of the *James Caird*, modified for a sea journey they were never designed for. This was unsurprising, given that the *Caird* was a lifeboat and not designed to cover vast distances under sail. In a bid to make things more workable, Nick had arranged the two ropes from the rudder so that they ran through wooden handrails in the front of the cockpit, the helmsman pulling them back toward himself to steer while braced at the rear. At least for landlubbers, pulling left moved the boat left and right, right—a godsend when the pressure was on and basic principles of sailing were forgotten under the stress of a threatening gray giant. But the lines required huge effort to pull and the friction of rope against rail added to the effort. Fingers fluctuated from feeling dead and numb to burning with pain as you adjusted your grip to get circulation back while your forearms ached with the effort. Each man chose the glove combination that worked best for him; I opted for woolen mitts as the rough wool helped me grip the rope. Still my fingers would

The mast cam shows the growing wrath of the Southern Ocean.

turn numb and white as the ferocious wind and wet mercilessly stripped warmth away and my old frostbite injury to my right thumb and forefinger came back. Nick fashioned a loop in the steering lines so they could be operated with the feet like stirrups. It was an attempt to reduce the burning fatigue in the arms and the damage to fingers, but only an experienced sailor could operate these deftly, so the rest of us persisted with painful, hand-based heaving to steer the boat.

Larso's face said it all as he came below to hand over the helm to Nick. He was a picture of sustained concentration, the frown lines deeper than normal and the ever-present smile replaced with a look of concern and focus. Conditions were worsening. "What's it like up there?" Ed asked. "Blowing dogs off chains," Larso grunted, throwing his sopping wet body into the pile of damp bodies on the floor in an attempt to force his way into some warmer place to get some sleep. Worsley's words again came to mind: "We were like those monkeys which, during a cold night in the forest, lock themselves into a ball for mutual warmth. If one gets left out and, pushing in, disturbs the others, a furious row ensues. So it was with us. When some shivering unfortunate on the outside tried to push in, there instantly arose a frightful burst of profanity and dire threats of vengeance from the disturbed men." Our reactions to the incoming helmsman never reached the same levels of disaffection, but our disgruntled mutterings were rough enough.

In the early evening half-light, I emerged on deck, my first bowel movement beckoning—perhaps at the prospect of my turn at the helm. I'd been considering the difficulty of the whole operation for the past hour, and was confident neither about how I would answer the call of nature nor how I would steer the boat. I squeezed down into the dark recesses of the cockpit by Nick's feet while he kept helm. The cockpit was home to torrents of icy water that rushed madly from one end to the other with the boat's wild movement. Listening for the metal bucket rather than trying to sight it, I fumbled in the dark, positioning it in a corner in the hope it might remain there. It didn't. The rodeo conditions sent it tumbling across the cockpit seconds later, Nick stopping it as if trapping a soccer ball. Returning it to its corner, I knelt in the cold water with nausea that had been kept at bay now rising fast with the need to focus on stripping back layers of wet clothing. With my trousers finally undone, I peeled off woolen layers down to the ankles, propping myself out of the water with one frozen hand as the other held the bucket just as a massive wave tipped torrents of icy water down my back and sent the bucket flying again. Angrily I regained it, tipping out the excess water, searching for a compostable bag and stretching it to failure over the bucket's sharp

Picture of focus: me during the storm.

rim, hoping it would not tear as it surely must. Hovering improbably over the bucket's wide rim (designed for accuracy, not load-bearing), my body seemed to sense the urgency and all was done in seconds before the next wave forced a brace in the corner of the cockpit to regroup. "All right down there?" shouted Nick mockingly. His humor evaporated as the bagged offering passed precariously close to his face before I tossed it overboard. "You picked a hell of a time," he joked as he vanished below deck to rest. I had picked the roughest conditions possible in which to christen the bucket and was duly proud of my achievement. The others still had the experience to look forward to.

Emerging on deck to steer in rough seas was like being in the passenger seat of a car, woken from deep sleep by the driver and handed the steering wheel as it entered a massive skid—there was very little time to readjust to the noise and speed at which the world around you moved. By the time you had focused and recalibrated, the hatch had closed and you were left alone on deck, responsible for everyone's safety. Alone, that is, save a skua or albatross wheeling high overhead.

I took the reins and began battling the bucking bronco, the *Alexandra Shackleton* reminding me of a small toy boat my sons play with. We were just a tiny speck in a vast ocean with the water crashing against the sides of the boat, making it shudder and groan with each big-wave impact. The top of our mast was

Ice forms on the sails.

routinely dwarfed by the waves around us as we lingered in dark, black valleys surrounded by malevolent walls of gray topped with hissing white foam. Each time waves above us threatened to curl over and engulf us completely, the *Alexandra Shackleton* would miraculously bob back up. I fought to keep her at right angles to the steep slopes of massive waves down which she careered, conscious that if I made a mistake, we would potentially be rolled. For a land-based explorer, it was intimidating and exhilarating in equal amounts. As Worsley said, "In a short time our ideas of size altered amusingly." Incredibly, being alone in the turmoil up on deck gave me the chance to contemplate the journey I'd been on personally up to this point and the enormous pressure I'd been under. The logistical, financial, and personnel issues that I'd dealt with right up until leaving Elephant Island had been intense, and, frankly, it had taken the enormity of this voyage to preoccupy me and banish thoughts of the issues that awaited me on my return. Survival in the Southern Ocean in the company of a team of great men was the perfect antidote for now. I remembered Discovery Channel's insistence that it was imperative for us to reach day four of our journey so they could get enough footage. It was a friendly but firm and oft-repeated maxim: without four days' worth of footage from the journey the documentary wouldn't be workable.

I was unsure of what day it actually was in our strange world. It was still day three — at least I thought it was, but I might have been wrong. Because we'd left at 7 P.M. from Elephant Island, new days theoretically came around only at 7 P.M. the next day, which was counterintuitive when daybreak normally symbolized a new day. We lived in a strange half-light below deck, our sleep constantly broken by the movement of people leaving or returning from watches, with every fifth hour signifying it was your turn again to battle the elements. And all this occurred as we headed toward an endless horizon with neither the evidence nor the means to show us whether we'd actually gone anywhere. Add to this days that lasted eighteen or nineteen hours, the absence of the sun meaning we didn't know our position, and the disorientation of using hundred-year-old gear, and we weren't sure exactly where we were, geographically or temporally. A big wave crashed over me, snapping me out of my fugue and reminding me of my immediate priorities.

Life below deck, meantime, was getting increasingly damp, with each man returning to the fold wet through. As his wet clothes touched you when he came off watch it only added to our group misery. Not that they differed much from your own clothes but, lying still, you'd deluded yourself that you were drier than you were.

Baz's military way of doing things was constantly frustrated by the lack of order down below. While he removed his jacket, folded it, and placed it where it could be found when next needed, it would always have been moved as a result of someone trying to find something of his own. However innovative we became about choosing unique spots for our gear, like wedging it behind a spar or in a corner, nocturnal migration always occurred. In the darkness below, to the right of the hatch, the jackets sat in a sopping wet heap as if they had fallen from a washing machine with a broken spin cycle. The only way to identify what was yours was by feeling for familiar contents in the front breast pocket.

Meanwhile, Baz, who had been nominated cook, battled to keep the stove going and everyone fed. And there were inherent dangers: every time the meths was lit to preheat the kerosene a huge orange flame issued forth, plus boiling water threatened to spill over Baz or us with every wave that hit. Other more insidious threats also lurked, in particular the feared hypoxia. I recalled Worsley describing the process of Crean lighting the Primus, keeping it balanced between his and Worsley's feet while it took three of them "to steady the Primus and cooker against the boat's violent motion." The hoosh that they prepared and ate "scaldingly hot" was a life-giving tonic to the men.

Even in the damp that was our world, fire was certainly a real risk. We lived atop 560 kilograms of camera batteries that were drip-fed by a battery recharging unit that ran on 40 liters of 99.9 percent pure methanol—one of the most poisonous and volatile substances around and one that should probably be avoided at all costs on a boat. Add to that each of us having sundry boxes of matches and lighters about our person and our full complement of eighteen flares, which if set off in a methanol-vapor-rich atmosphere would act like a detonator and, as Seb delicately put it, "Then you can kiss your arse good-bye."

For us it was more crowded on board than it was for Shackleton and his men. Our batteries for filming at least added little to the space constraints since we simply slept on them rather than rock ballast as Shackleton had. I can report, however, that neither was conducive to a decent night's sleep. One could only imagine the seriousness of the crime for which we were doing penance as we lay in our itchy wet hairshirts atop the solid, angular surface of batteries and rocks. Extra items we carried meanwhile included six cameras for filming, which protruded from various corners of the boat (two below deck, four above); a large control panel that enabled Ed to change over memory cards and view footage he had taken; four ten-liter plastic containers containing pure methanol for our battery recharging unit; the unit itself, which sat in a dedicated protective wooden box about the size of an old-fashioned TV set; six lifejackets; a VHF radio; and six large duffel bags, each containing an immersion suit and a dry suit. That combined with the greater physical size of us as a crew—Tom Crean, whom the navy records as being five-foot-ten (although Worsley claimed he was over six foot), was referred to as *Endurance*'s "Irish giant," whereas our shortest man, Nick, was five-foot-eleven with me standing six-foot-five—meant there was scarcely room to move below deck.

The most dangerous section of the boat was the fifteen centimeters that separated the below-deck access hatch and the cockpit, which each man needed to sit astride momentarily when going on or coming off watch. It felt high and exposed, particularly in a big sea when the *Alexandra Shackleton* was perched atop a massive wave as you looked down into deep, forbidding valleys either side of its crest. All of us had close shaves where a sideways impact from a wave threatened to knock us either through the hatch, into the cockpit or, worse still, overboard. Nevertheless, we became adept at balancing with our legs and holding on to the runners on the deck placed there for that very purpose.

I recalled Thor Heyerdahl's balsa-wood log-raft journey across the Pacific aboard the *Kon-Tiki*. He described saving one of the men who fell overboard by throwing a rope to him as his bobbing head and flailing arms quickly receded into the distance with the current. We had a modern throwing line or Jonbuoy (which for me conjured up ridiculous images of the eldest son in TV's *The Waltons*) and a life jacket if we chose to use it. We had also considered putting out twenty meters of knotted rope behind the boat so that a man might grab it, but the chances of his being able to do so once in the water were virtually zero. As it was for Heyerdahl, if anyone went overboard, it would be curtains unless the Jonbuoy could be thrown because, without a keel, we had no chance of turning around to pick him up.

Nightfall for us was defined as that time when you could no longer read the compass without a candle, particularly as the compass sat in a recess under the lip of the cockpit. It was a lovely old piece of maritime history—heavy brass weighted with lead and oil and gimbled to allow it to move, in theory at least, independent of the motion of the boat. Nighttime was Larso and Nick's time because they could steer more accurately than the rest of us without the compass, using instead the sound of the wind in the sail, the feel of the boat, and the flap of the pennant, just visible in the wan moonlight on the mizzen near our heads. But they still needed to occasionally check that the wind hadn't shifted and sent us off course.

The conditions incessantly frustrating us made even the simplest tasks difficult. Sometime during what we agreed was our fourth night, Larso shouted down for a candle. Baz and I found one and began searching for dry matches in tins in our pockets with which to light it. After seven or eight attempts we finally did so, sheltering it as best we could, but as we exited the hatch the candle went out repeatedly, each failed attempt being met with increasing frustration. Baz climbed into the cockpit and tried several more times. The candle could not, however, be lit in its position in the glass case that housed it in the rear of the compass due to the tight space, while the "still" air by Larso's feet was too drafty to allow us to light it there. Baz removed the glass housing from the back of the compass and returned below deck with it and the candle, cursing under his breath. By now, fifteen minutes into our attempt, our matches had stubbornly decided not to work. We woke Ed for a lighter after Seb fruitlessly searched for our spares bag in the darkness. With Ed's lighter the candle lit the first time and was placed carefully into its housing for the almost ritualistic journey back to the compass. We felt like Neanderthals carrying a burning log from their first fire, fearful of its going out. The air gap in the housing that let in enough air to keep the candle alight, this time let in too much. The candle extinguished again. Baz finally took the candle, lit it by Larso's feet with the lighter, and swiftly put it back into its protective housing. After all our efforts, Larso checked the compass and, reading "North," put the candle out immediately to conserve it, just as the *James Caird* crew did the few times they used their only candle, saving it, as Worsley recalled, "for emergencies—principally making the coast."

The frequency of loud booming hits from waves to the side of the boat was abating, and it seemed that the wind that had raged incessantly for the past two days was dropping. Larso was at the helm. As ever, I marveled at his and Nick's ability to read the wind. Suddenly the VHF radio crackled into life, a sense of urgency evident in Ben's voice. "We can't see you," he said. Given his serious tone, my humorous response—"We can't see you either"—might have been misplaced. "We're not picking up your AIS or radar," he continued. I wasn't surprised at the radar not working—a small wooden boat with linen sails like ours would provide little for it to get a reflection off—but the Automatic Identification System not working meant not only could *Australis* not track us but also that any other vessel in the vicinity would not see us either. Although few ships ply these waters, it would be an unfortunate irony to be run over by an "In Shackleton's Footsteps" icebreaker or tourist vessel heading to South Georgia from Elephant Island. Sure enough, the telltale lights behind the wooden casing that normally indicated that the AIS was on were dark and lifeless. We told Ben we'd investigate immediately. Even though the wind was dropping, the seas were still huge from the storm. With darkness upon us our small, unlit vessel would be impossible to see even in moonlight.

Seb clung to the deck, skittering dangerously in his wet leather boots as he tied our large cooking pot to the halyard and hoisted it to the top of the mast. We radioed *Australis* to see if the largest metal object we possessed, atop the tallest structure we had, was visible to their radar, but the VHF refused to function either, courtesy of the same power failure that had killed the AIS. Again we borrowed Ed's lighter to light the oil lamp in order to find the spares bag that contained the battery-operated VHF. This too was dead due to its getting wet, and a repository of spare batteries for it lay in a tin hidden somewhere else. Once we had tracked this down by candlelight, the battery-operated VHF was switched on and we contacted *Australis*. Ben confirmed that the radar was performing about as well as our waterproof jackets and that we were officially radar invisible. With our battery VHF becoming increasingly strained by distance from *Australis*, I precariously clambered up on deck and positioned our emergency light high on the mizzen mast so that at least *Australis* could follow us while we sorted out our problems. In keeping with the way things were going, the

Nick: along with Larso, our most accurate sailor.

light also decided to malfunction as waves continued to periodically drench us. I dried it out below by the glow of the oil lamp and returned it to its position at the top of the mizzen mast, where this time it worked. Finally *Australis* confirmed they could at least see us, which, although an improvement, without the AIS would be completely impractical in bigger seas, or mist. More than ever we felt very alone.

After multiple attempts, Seb still couldn't get the electrics to work, although Ed's cameras continued to function. We sat huddled by the light of the oil lamp, thinking through our options. Each life jacket contained an AIS of its own with a theoretical life of twenty-four hours and apparently a four-kilometer range. If we couldn't get the main AIS going, we could run each life jacket AIS in sequence throughout each nightly period of perhaps eight hours of darkness to enable *Australis* to track us. The computer screen would register one of our names rather than the boat but, apart from that, it was a good solution. Six life jackets should give us a total of eighteen nights' coverage in the worst-case scenario. Frankly, if we weren't at South Georgia by then, the AIS would be the least of our problems. With great anticipation we switched one on but *Australis*, keeping us in sight via our rear light, could only pick up the signal when they came within 400 meters of us. And that was with the life jacket AIS on deck, far higher than a man would be flailing in the dark water. It didn't say much for the chances of either boat or individual should the worst happen.

Their heads are green, and their hands are blue,
And they went to sea in a Sieve.

Edward Lear, *The Jumblies*

On a journey using period clothing and equipment, it seemed ironic that the six of us were sitting cramped below deck on a hundred-year-old whaler out in a dark, stormy Southern Ocean trying to fix twenty-first-century technology issues. Having been closely involved in the system design, Seb got straight on the case trying to think through what could be wrong while the rest of us "helped" him via constructive interrogation designed to spark a flash of inspiration and, better still, electricity. The cameras and the communications gear were powered by two banks of batteries separated by a voltage-sensing relay to protect the system from overcharging. As the cameras still worked, the problem could be something to do with this or the fact that in the storm we had allowed our battery-recharging unit to run out of methanol. Or perhaps the camera batteries had retained more charge and not yet run out. With our brains fuddled by lack of sleep, darkness, boat movement, and nausea, we decided to wait until daylight and, hopefully, calmer weather. We just had to hope *Australis* could follow a small light on a tiny boat in big waves on an enormous ocean.

We had to get the system up and running again because without it our ability to complete the expedition independently would be in serious doubt. If *Australis* couldn't physically see us or pick us up on radar or AIS, we would have to abandon our bid to be the first to sail a replica *James Caird* using only traditional navigation, needing to resort to modern means to let *Australis* know where we were. It would be a crushing blow. The only upside was that we now knew how much power our batteries and AIS were consuming by how quickly our methanol cell ran dry putting power back into the system.

The Alexandra Shackleton *heeled over in the mist.*

Previous pages: Approaching the towering cliffs of South Georgia, one of the most dangerous parts of our journey.

Our new appreciation of the range limit of the life jacket AIS units also gave us some serious food for thought. Larso had gone on watch in darkness swearing under his breath about how much of a pain in the backside they were anyway. In so doing he voiced what all of us felt. Pulling them on over a bulky, wet, gabardine jacket, straps always snagging on its buttons, while balancing on your knees in a crowded, rocking boat, head stooped low with no peripheral vision and no one to help was at best challenging and, in darkness, next to impossible. Also, due to our differing builds, we each had our own preadjusted jacket, but, as with all our gear, they all looked identical from a distance. The chances of grabbing our own jacket first go was 1:6 and those odds bore out much of the time. If someone had gone overboard and the AIS had activated, those on board *Australis* would have been perfectly entitled to wonder who was actually in the water.

Regardless of the danger, we decided not to use our life jackets, as the weather appeared to be marginally improving. Nick had some misgivings about aspects of their preattached safety lines anyway, in particular the fact that the longer of the two lines on each jacket was too long and got in the way, while in a capsize the shorter safety line might prevent the helmsman from finding the air pocket in the top of the cockpit in the upturned boat, attached as it was to the deck.

The wind was lessening, the seas were beginning to peak, and fewer waves broke into the boat. But snow fell heavily throughout the night, coating each helmsman with a layer several centimeters thick on the windward side of his head. The sun momentarily revealed itself through the breaking clouds for the first time since we'd left Elephant Island only to sink spectacularly, large and red, into the sea, cloaking us in darkness with, eventually, just a handful of stars bright enough to shine through. Several hours after the sun sank beneath the waves, the moon appeared, its yellow light shimmering magically on the black, uneven surface of the sea. For a group of tired men whose nerves had become frayed during the previous forty-eight hours, it was a godsend and reminded us why we were out here.

Suddenly I was struck by the differences between a storm on land and one at sea. I had spent five months exploring the desolate interior of Antarctica on foot, pulling a sled all the way to the South Pole and beyond, and living in a tent high on the windy polar plateau. The winds there blew with a force that easily rivaled or eclipsed what we had experienced here on the sea. The key difference, however, was that the land surface didn't change as storms raged above you. Here on the ocean you got it from both sides—from the elements above as well as the

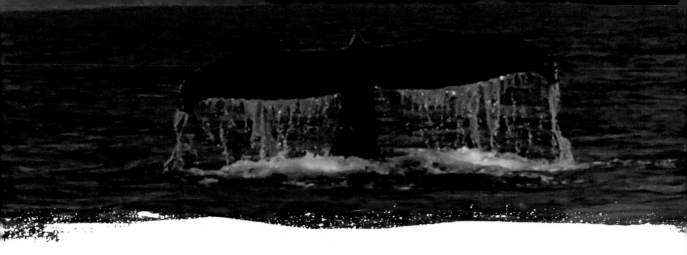

Every now and then we were reminded that what was an alien ocean to us was home to many.

surface over which you traveled that changed unrecognizably. Here there was literally nowhere to hide.

Larso and I were on deck in the dawn light when a loud blowing startled us both. We turned to see a massive glistening form appear only an arm's length away, gliding effortlessly through the water and virtually brushing the side of the *Alexandra Shackleton*. We could even see the barnacles glistening on the back of this southern right whale as it surfaced, its huge form rotating to see us better with one eye out of the water. There was a certain irony in having a whale surface to satisfy its curiosity about these men aboard an old whaling boat. It certainly wouldn't have done so a hundred years ago, or even fifty for that matter, when those men so admired by Shackleton were still decimating whale populations. Even this whale's name was indicative of its value to man. They were called the "right" whale because they contained a lot of oil and didn't sink when harpooned, which made their retrieval easy. This thrillingly intimate encounter with this giant of the deep was wonderful. It felt as though the whale might have popped up to ask, "You made it through that storm all right, then?"

The cold night turned into a cold, crisp early morning with a light frost forming on the *Alexandra Shackleton*. We weren't, however, experiencing as cold a journey as Shackleton had by this point. For him a treacherous coat of ice developed on the *Caird*, increasing in thickness as each day passed. By day eight the ice had caused the loss of two of their four oars and threatened to capsize the boat due to its increased and uneven weight. As Worsley recalled,

Something had to be done, and quickly, so we took it in turns to crawl out with an axe and chop off the ice. What a job! The boat leaped and kicked like a mad mule; she was covered fifteen inches deep in a casing of ice like a turtle-back, with slush all over where the last sea was freezing.

It must have seemed like a silent, tenacious predator circling and then smothering its prey.

Thankfully, and in no small part due to the fact that our journey took place three months earlier than Shackleton's, we were not affected by ice accumulation. On the land-based polar trips I have done, however, it is fluctuating temperatures and wetness that so often cause the worst problems. Snow melting and then refreezing in wet, windy conditions can quickly result in the onset of hypothermia. That was what we were experiencing now. Sometimes in extreme cold I've known ice to form on clothing in a way that effectively seals it and prevents wind from getting through. But as we weren't likely to experience freezing, dry conditions, we couldn't expect such a positive side effect.

By late morning of day four the sea was improbably calm and the temperature the warmest of the trip. The previous night's red-sky sunset had delivered what it promised. By noon, the anger of the sea during the previous two days had completely vanished and it was now like a millpond, our doubts and difficulties as hard to recall now as the sea that had caused them. Wasn't this always the way? Good and bad times come and go, but while the former reign, it's as if the latter never existed. The reverse applies just the same, and in the throes of that storm, we had thought it would never end.

One thing we knew for sure, the calm wouldn't last, so the boat became a hive of activity above and below deck, with Seb taking advantage of the calm and light

Improbably calm: our concern of the previous days now seemed difficult to recall.

to revisit our power issue. His final take on things was that we should remove the two batteries in the bow from the circuit and reconnect a new wiring loom by literally twisting the exposed wires together to short-circuit the system back to life. This was met with some understandable trepidation from Ed, who had visions of his camera bank going up in smoke. Seb, however, was confident that no harm would come to the camera battery bank, and on that basis I told him to go ahead.

With predictable difficulty, I dug out a new methanol container from the bowels of the boat and connected it to the system. I followed the instructions on the last page of the troubleshooting section that more or less said: "If you're reading this you've probably made the unforgivable mistake of allowing the system to run out of methanol." I consoled myself with the fact that the average person wouldn't likely be using one of these systems in the tiny confines of a hundred-year-old boat in a storm down in the Furious Fifties. But it wasn't going to solve the problem.

Having connected our respective parts of the system, we were satisfied we had done all we could and sat back to take stock. In a scene vaguely reminiscent of a thriller where the pliers hover over the green and red wires of a bomb before one is cut, we switched the system on, hoping in our case for light, not dark. After a couple of seconds' delay, the AIS panel and VHF lights came to life, followed by the cameras. Ben confirmed he could again "see" us and calm returned on all fronts. Overcoming this latest problem gave us tremendous confidence and had forced us to understand more about our systems into the bargain. Add to this the fact that the *Alexandra Shackleton* had survived the storm well, and we had some great positives to build on.

By midmorning the sun had melted the accumulated snow on the deck and it became a Chinese laundry of wet reindeer skins and sodden boots. As our little boat sat becalmed in the doldrums beneath a blue sky, Baz cooked below while the rest of us caught up on sleep or tried to track down hats and gloves that had been lost in the bilges since we left Elephant Island. So debilitating and difficult had the conditions been until now that most "nonessential" activities—such as eating, drinking, and attending to personal hygiene—had fallen by the wayside. Nick's log entry read: "Zero wind. Full spring clean. All fed and watered. Baz cooked. Clothes dried. Larso sunburnt. All happy. Some personal space re-established. Teeth brushed." Along with describing the conditions, Larso's log declared, "Farting a good sign." It was a serious point. Up until now no one had eaten enough to generate gas.

In the meantime, there was another source of gas on board. Worryingly our port-side water barrel, one of the two old whisky barrels that we'd filled at Arctowski with glacial water, smelled very eggy and had, we decided, become contaminated. How it had happened was a mystery. This unwelcome development provided another extraordinary, unplanned parallel between our journey and Shackleton's. One of his water casks was stove in slightly when the *Stancomb Wills*, which was towing it to the *James Caird* ready for their departure, was driven by the swell onto some rocks, rendering the water within "brackish." This proved to be a serious problem as it led to the crew being dangerously short of water later in their journey. We needed to keep using the water from our affected barrel, but planned to boil it for use in meals and hot drinks rather than drink it cold. However, the combination of our ever-malfunctioning Primus and our impatience meant, I suspect, that we often stopped short of letting the water reach true boiling point. Since most of us already felt slightly unwell, it was difficult to tell if we suffered any ill effects from the water.

The Primus stove on the *James Caird*, as with us, played a vital role in maintaining both energy levels and morale. In Worsley's words, "It was [Shackleton's] principle to fight the cold, and constant soaking, by ample hot food." As a result, he recalled, "Sir Ernest had the Primus going day and night as long as we could stand the fumes, then it would be put out for an hour. This and a generous drink of life-giving hot milk every four hours, at the relief of the watches, kept all hands from any ill effects." We too used our stove as often as we could, with Baz being the Crean of our group in terms of both his cooking and his uncomplaining, stoic nature. Even though he probably suffered the worst seasickness among us, he was always ready to help others.

After a brief visit from the Raw TV crew by Zodiac to collect some film, the next priority for us was working out where we actually were. We could only really make an educated guess based on dead reckoning of our speed and direction of travel. In order to get a decent sun sight, the disc of the sun needed to be visible in the sextant, but now a bank of high clouds was rolling in from the west and we knew the clear skies would likely be gone by noon Greenwich Mean Time, when sightings could be most easily calculated. Worsley's poetic words yet again seemed to describe our conditions, with the exception of the past few hours of blue skies: "Never a gleam of sun or stars showed through the dull gray or else storm-driven pall of clouds that in these latitudes seems ceaselessly and miserably to shroud the bright blue sky and the cheerful light of sun or moon." This was one

Top: The on-board laundry. Trying to dry our jackets on deck.

Bottom: Calmer weather draws small smiles from Nick and Baz.

of the most difficult aspects of being out here—we were constantly at the mercy of the elements, not knowing what weather was coming, nor in which direction it would allow us to travel and how fast or safely.

In the meantime, while it was great that the sun was drying our possessions on deck, we weren't getting anywhere. The sails sat listless and our positive mood deflated as the clouds slowly moved in. I realized just how important it was in a project of this enormity, and involving a team of this caliber, to feel that we were constantly taking positive steps toward our target. I realized too how much I had begun to rely on the momentum that can be sailing's reward. It moves you toward your objective even as you sleep—complete anathema to the sled pulling I was used to on expeditions, as a sled progresses only when you pull it. While you sleep you go nowhere; in fact, on a journey to the North Pole, you often go backward while sleeping with the southerly drift of the ice. I chided myself for slipping into the mind-set where I expected to get something for nothing and resolved to focus instead on our achievements to date and the positives of our current circumstances. We were drier than we had been for days and the sun was about to tell us where we were, if the clouds could only keep away for just a bit longer. Plus, as dangerous and uncomfortable as it had been, the storm had carried us hundreds of kilometers toward South Georgia.

When 3:30 P.M. finally came around (noon GMT for our approximate position), banks of clouds had settled over us and only a pale disc of the sun was visible high above. Larso steadied himself on deck and began measuring the angle to it with the sextant at two- or three-minute intervals prior to and after 3:30 P.M. He needed to make sure we got the sun at its highest point, shouting "Mark!" each time. Never one to miss a chance to joke, Baz quipped, "I'm happy to be timekeeper but can you stop calling me Mark," as he recorded the precise time of Larso's pronouncements. Seb's job was to offer a corrected time based on how many seconds we knew the chronometer was losing each day—about four seconds, enough to leave us way off target by the end of the voyage. Satisfied, Larso joined Nick below to create some navigational alchemy as the rest of us looked on expectantly, awaiting their verdict on how far we'd come. "Four hundred fifty miles, give or take," announced Nick, ". . . to go, that is." We were ecstatic: it meant we'd covered 300 nautical miles in only four days, which had included periods of very light winds as well as stormy periods when we'd been forced to trim all of our sails.

There were many reasons to be happy with our position. Not only were we successfully keeping north of a direct bearing to South Georgia, giving us more

Celebrating our first sight of the noonday sun with a few drams of Mackinlay's.

leeway later should we be pushed east by the westerlies that prevailed here, but we were also making good progress toward the 54°S line of latitude at which point we could turn east and head in to South Georgia. Furthermore, we were safely away from the pack ice and the danger of being drawn toward the South Orkneys. Last but not least, we were well ahead—perhaps 100 nautical miles—of Shackleton at the same point in his voyage.

Frustratingly, however, on day six strong northerly winds took the place of the stormy southwesterlies we'd had, making it hard for us to consolidate our northward position. And the awkward, mixed sea state resulting from this change of wind created big waves that struck us side-on. Combined with incessant rain and bitter cold, these meant we reached a new level of soaked-to-the-bone misery. It was ironic that the southerly winds that blew directly from the Antarctic were not as cold as the northerlies we were now experiencing.

Northerly winds don't, of course, mean you can't gain ground to the north, particularly in modern sailing boats that can travel at a forty-five-degree angle into the wind, their keel preventing them from yawing sideways or backward as they do so. The absence of a keel on the *Alexandra Shackleton*, however, meant we could only manage just fractionally to the north of due east in northerly winds.

Nick and Larso shared between them the darkest hours, in great feats of focus, while the rest of us did the dawn, dusk, and daylight watches as they slept. Several of us noticed that, with a bit of effort and concentration, we could steer the boat in a more northerly direction into the wind, almost to east northeast. While we were pleased with our achievements, Larso explained in no uncertain terms that this was a false economy and we needed to stick to our instructions. He explained that the slower speed meant what we thought was better progress to northward in fact resulted in us being pushed further sideways—in this case to the southeast—by the wind and current. Initiative was to be encouraged, but not if it got the wrong result!

Little escaped Nick and Larso. Even when Larso was meant to be fast asleep below deck, any erring too far north that caused the sail to flap or, God forbid, to jibe, would elicit a frustrated cry of "Jibe!" On one or two occasions, he appeared back on deck to provide some friendly advice to the helmsman as to what it was we were actually trying to accomplish and the importance of not deviating too far from it. Larso, in his typically understated fashion, later recorded our efforts in the northerly winds in his log: "Sailed with full sail and could steer North of East using only leeward steering line to steer with. The luffing mainsail

A soaking wet Ed at the helm forces a smile.

was our reference to steer to. Wind built through night with snow and rain. As seas built some water coming over deck. A couple of random gybes but generally everyone steered the course well."

Advice on what to do on board—and when and how to best achieve it—was normally met with good humor by the nonsailors, although some terse exchanges did occasionally occur. More than once I had to tell everyone to calm down and focus on our goal, but really it was just a matter of getting used to one another's ways and nothing more.

In reality we were a team of people who were used to being in charge. But we each brought a unique skill set to the table without which the expedition wouldn't be possible. Certainly with every passing day on board the *Alexandra Shackleton*, I realized it was Nick and Larso's wealth of sailing experience that were edging this little boat toward its target and in a way that was making a very dangerous and challenging journey seem manageable. And not to do the rest of the crew a disservice, I was also impressed with how instructions were taken in the spirit in which they were intended and the focus given to the task of helming accurately on what was a very steep learning curve. Before the voyage Seb had placed a plaque in the boat that quoted Horatio Nelson at the Battle of Trafalgar: "England expects that every man will do his duty." Without doubt, our small crew was living up to this in every respect, and I was proud just to be part of this great team of men.

The problem with the easterly direction we'd been forced to travel in for the day of northerlies was that it made us far less confident of what bearing we'd managed to keep. We hoped it was a bit north of east but, with the current causing us to drift inexorably east at half a knot an hour and the wind pushing us south, we really weren't sure. Now, just as we had learned to resist the *Alexandra Shackleton*'s desire to jibe, the wind blew in strongly from the south southwest. It was a better wind direction for traveling north but again resulted in confused seas as waves created by northerly winds were pushed upright into menacing gray walls by winds now blowing from the opposite direction. Although the *Alexandra Shackleton* continued to bob like a cork over much of what, at the beginning of our journey, I would have assumed would sink us, the confused seas still periodically sent great slugs of water into the cockpit and over the helmsman. In seconds such an inundation undid five hours of painstaking drying of your clothes using your body heat down below.

A sudden change of wind direction suggested something big had arm-wrestled the northerlies into submission and was ominously coming up behind us from the southwest. A towering turquoise iceberg—the first ice we'd seen for five days—added to the sense of disquiet on board and did nothing to allay our fears that we might have drifted a lot farther to the east toward the pack than we'd hoped. That night Larso's log recorded: "'Sailing downwind again therefore heavy and difficult steering. Compass light kept failing so ended up going 'old school' and just using rear masthead streamer to steer by. Downstairs very uncomfortable." He went on to say, "A gray day. The large swell that was rolling through—something in the Drake Passage was causing this." It was certainly impressive to see the awesome power of nature at work with these massive rollers coming through. We just hoped they didn't herald something too sinister heading our way.

In terms of life on board, we'd adjusted as well as we could to six men sharing a double bed. Ed and Seb had moved their sleeping positions and now lay on top of the food provisions, cordage, and spares at the front of the boat. As Seb was at pains to tell me, none of this equipment was lashed down—had the boat rolled over, he and Ed would have been pinned by it against the boat's ceiling. It would have been a hell of a way to go—smothered by safety gear that was designed to save life, not shorten it—but their repositioning made things more bearable for the rest of us.

As far as toilet duties went, we urinated into a bottle below deck and it was very poor form to use it and not dispose of your contribution immediately lest another man find himself presented with a full bottle. Sometimes, of course, it was impossible to extricate oneself from the tangle of bodies below deck to do this—at least that was each perpetrator's defense. Bowel movements were conducted perched over the metal bucket in the cockpit by the outgoing helmsman's feet or, if you were feeling adventurous, alone while gripping the two steering ropes with one hand. To ensure a degree of privacy and to save the film editor some nasty shocks we would hang our beanies over the 360-degree infrared camera in the cockpit that otherwise captured everything.

Shouts of "Berg!" woke Ed from his slumber to get some footage. A gifted cameraman and one of the hardest working around, like an old gunslinger he seemed to sleep with one eye open, ready to spring into action if there was something worth filming. Getting him on deck from his new forward position, however, was an awkward proposition. He and Seb lay side by side like Tokyo businessmen in a rent-by-the-hour pigeonhole hotel, the ceiling thirty centimeters from their faces. In a bid to get footage of the iceberg, Ed had to be posted out horizontally on a sea of hands like a crowd surfer. Near the hatch he placed his

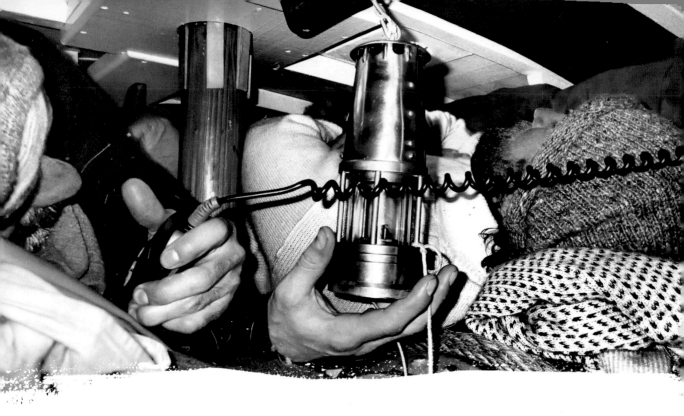

Close encounters: Ed and Nick down below. From this position, Ed would have to be posted toward the hatch horizontally, on a sea of hands.

knees to his chest in a tuck position before standing up and poking his head out.

There certainly were some magical moments on the high sea. That night I became conscious of shapes all around me as I helmed alone by the light of the moon. With stars visible for virtually the first time during the voyage, black-and-white shapes flashed below the boat from one side to the other. They were hourglass dolphins and were clearly curious to see anyone else out here. Nick's watch followed mine on what was becoming a very cold night. The south south-westerly winds again required navigating by the wind in the sails in the darkest parts of the night with occasional checks via candlelight. Nick recorded the ominous words "Toes still numb" in the log, having lost sensation in them since day three—something that would come back to haunt him later.

Never wanting to release its grip on us too easily, the Southern Ocean gave with one hand and took with the other, the clear night being replaced by dense fog and rain that, combined with a fickle following wind, made for depressing and difficult travel. The sail always threatened to jibe if anyone other than Nick or Larso was at the helm and momentarily lost concentration. Each helmsman was wet through again and again with rain and wave action, and we had long since given up any hope of drying out. Abandoning the thought seemed to help psychologically.

Perpetually wet, each man chose his own regimen for managing his boots and clothing. Seb, Larso, and Nick wore long leather jackboots, Ed and Baz old leather ankle-high boots, and I the old military hobnails I had worn on my Mawson expedition in 2006. While Nick and Ed were suffering worst from numb feet, the rest of us were not far behind. My boots were well and truly broken in but the predrilled hobnail holes provided a route to let water deep into the fabric of the boots, turning the leather into pulp. As a result and through habit from years of polar expeditioning, I removed them at the end of most of my sessions at the helm to allow my feet to breath and warm up. Baz religiously did the same from the outset, putting on his one pair of dry socks as he slept and advising others to follow the same regimen.

The sun made a brief appearance the following day but not at noon, so it was difficult to calculate our exact position. Nevertheless, in a rocking boat and with the wind again having changed direction, this time to northwesterly, Larso took several sights in big, confused seas. He and Nick calculated that we had 126 miles to run. They weren't at all confident of these non-noon sights taken in rolling seas, but at least they seemed to tally with our dead reckoning that suggested we had less than 150 miles to go.

This dense pea-souper fog made us wonder, however, what it would be like if we had similar conditions in a few days just off South Georgia and the mood on board became strangely subdued at the prospect. Up until now, all of our energy and focus had been on trying to reach South Georgia. Now we began to consider how we might avoid it if visibility was zero and a wind was blowing us toward the rocks.

Visibility was now only a few hundred meters and, although it was obviously an optical illusion, I was convinced we were heading steeply downhill toward the edge of some gray abyss for the whole two hours of my watch. We'd only been out here for ten days but we were desperately looking forward to terra firma to escape the wet discomfort and uncertainty of life on board. I felt as if I'd been out here for months.

Larso was on watch that night and, although the winds were not big by the standards we'd become accustomed to, the changed direction from which they now came was producing some big waves in a mixed-sea state. Without warning, a huge wave hit us, causing the boat to lift violently and shudder as an overwhelming wall of water crashed its full weight into the cockpit. "What was that?!" Larso shouted as Ed on backup below began bailing strenuously. It wasn't a wave of the caliber of the rogue that Shackleton experienced and described so powerfully in *South*, but it was menacing nevertheless:

> It was a mighty upheaval of the ocean, a thing quite apart from the big white-capped seas that had been our tireless enemies for many days. I shouted "For God's sake, hold on! It's got us." Then came a moment of suspense that seemed drawn out into hours. White surged the foam of the breaking sea around us. We felt our boat lifted and flung forward like a cork in breaking surf. We were in a seething chaos of tortured water; but somehow the boat lived through it, half full of water, sagging to the dead weight and shuddering under the blow. We baled [sic] with the energy of men fighting for life, flinging the water over the sides with every receptacle that came to our hands, and after ten minutes of uncertainty we felt the boat renew her life beneath us.

Thankfully we weren't visited by a wave of that size again, although several threatened, only to pass harmlessly beneath us. Meanwhile the northwest wind of the past twenty-four hours had pushed us toward South Georgia at a good clip— perhaps three to three and a half knots—meaning we were likely within fifty miles of South Georgia if our previous position was at all accurate. Finally, in an act of

Paul running the numbers from our sun sighting.

Day	Lat (Estimated Depart.)	Long	Lat (Fixed)	Long	Diff	Dist		Notes
1								
2			60°49.096	54°58.727				
2	66°00 s	54°40 w	60°11.786 s	54°19.702 w	11 nw	64	64	EP
3	– s	– w	59°40.2 s	53°21.0 w				
3	58°45 s	52°50 w	58°53 s	52°51 w	8 s	90	145	EP
4	– s	– w	58°17.7 s	52°23.1 w				
4	57°43 s	51°25 w	57°37.90 s	51°37.011 w	8 se	84	229	EP
5	– s	– w	57°07.891 s	50°54.082 w				
5	56°52 s	50°37 w	56°50.010 s	50°46.744 w	5 e	55	284	15305ITE
6	– s	/ w	57°00.300 s	49°33.407 w				
6	56°41 o	47°25 w	57°09.90	48°07.1 w	36 sw	88	313	EP
7			57°03.855	47°30.62 w				
7	56°06 s	46°20 w	56°39.113	46°39.33 w	35 w	56	367	EP
8			56°14.99 s	45°34.50 w				
8	55°25 s	44°17 w	55°52.0 s	44°51.7 w	33 ne	75	444	EP
9			55°20.3 s	45°40.9 w				
9	54°52 s	42°50 w (1730)	54°51.5 s	42°36.3	45	97	540	EP
9	54°28'	41°30						
10	54°20'	39°40 (1030)	54°32.1 s	40°40.2 w	37		597	
10	54°12'	38°44'	54°23.00 s	39°51.4 w	40	97	626	
11	54°17 (0930) s	38°53 w	54°10.0 s	38°39.0 w	10		666	
11	54°16 (2000)	38°26						
2								

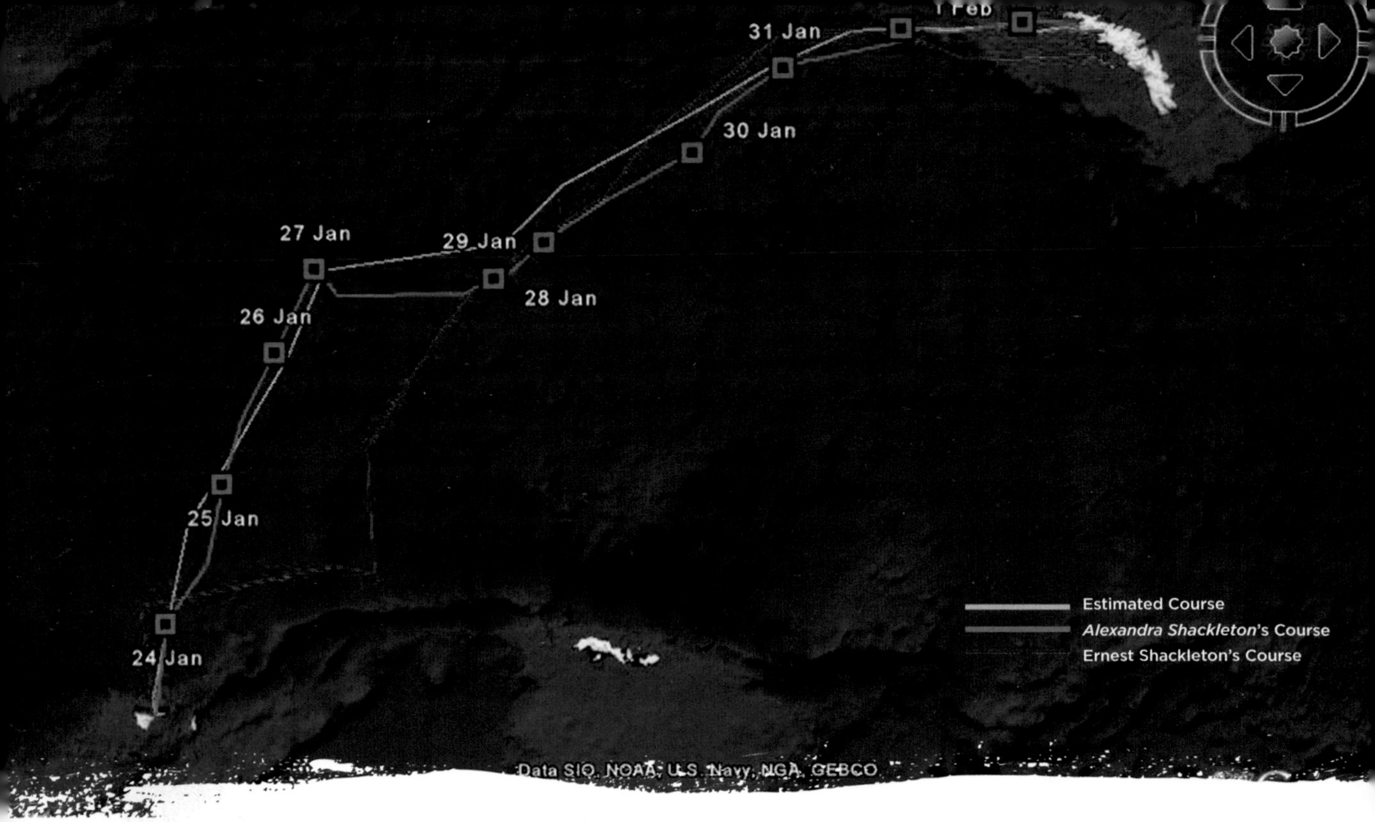
1 Feb
31 Jan
30 Jan
29 Jan
27 Jan
28 Jan
26 Jan
25 Jan
24 Jan
Estimated Course
Alexandra Shackleton's Course
Ernest Shackleton's Course
Data SIO, NOAA, U.S. Navy, NGA, GEBCO

Shackleton's route, where we thought we were, and our actual route.

charity the following day, the gods granted us clear weather and we were able to get a noon sight, only the second of the voyage. It seemed to indicate we were fractionally to the north of the latitude of King Haakon Bay, out some forty-three or forty-four miles to the west. If this were right, it meant that a northwest wind represented a good angle of approach for us. If it held, it would allow us to bear away and head back out to sea if we got our approach angle for King Haakon Bay wrong, something that a wind directly from the west or south would not allow us to do as we would find ourselves in a lee shore situation, particularly given the shape of South Georgia's west coast. Like a crescent moon orientated diagonally from northwest to southeast, its ends curved inward in a 145-kilometer-long line of treacherous rocks, cliffs, and glaciers, with bays offering some shelter. It was a big target to aim at now we were approaching from the west, and the prospect of finally seeing our goal for the first time was exciting. We expectantly waited for it to emerge out of the mist like a lost world.

The gray descended immediately after our noon sightings had been taken, taunting us by showing us how close we were but refusing to reveal our prize. Shackleton had known land was close by when he saw kelp and birds like shags and terns, which never venture far from land. Regardless of how much we willed the albatrosses and skuas that had accompanied us for much of our journey to morph into their less graceful cousins and prove our proximity to land, they did not. All we got was a Chinstrap penguin noisily and repeatedly cawing to Larso, "Nor Nor," to which Larso responded, "South South." "I don't care what he says," said Larso under his breath, "I'm heading southeast." It was definitely time to get off this boat.

That night we decided to take the mizzen down and sail with just the jib and main, shaving perhaps half a knot of boat speed in a bid to reduce the chances of smashing straight into our target in the darkness and mist. We could be as little as five miles away or as much as twenty-five, yet still South Georgia refused to reveal herself. Somewhere in the mist was our curious combination of savior and nemesis, rendered all the more tantalizing by our disbelief that we could have possibly made it using only dead reckoning and two sets of noon sights.

The radio crackled to life in an unscheduled contact from *Australis*. "How much longer are you intending to go for tonight?" Ben asked innocently. We replied that we were intending to go for another hour and a half, up until 11:30 P.M. or so. Then we'd put out our sea anchor and sit off what we hoped would be a mountainous island somewhere in the fog bank. There was a pause at his end. "I suggest you do this now," he said, giving no more information. Not wanting to ask him questions to which we didn't want, nor could he give, an answer, we did so immediately. Immobilized by the sea anchor, we discussed what his comment might mean. We had briefed Ben beforehand that we only wanted to hear from him if an unavoidable danger lurked that we were sailing into blind, or if we had gone beyond South Georgia with little chance of sailing back upwind. Could we have gone past the point of no return past South Georgia, or were we about to sail right into the island? If the former were the case, then the experiment of doing things unsupported was likely over unless northeasterly or easterly winds blew us back to the island from our position beyond it out in the open Atlantic. This was highly improbable—more likely we were closer than we thought and were heading into it.

I took the first watch of the night from midnight to 3 A.M. as small bits of kelp did now start to occasionally drift past as we sat stationary on the sea anchor. I read Worsley's account of the Shackleton boat journey by candlelight, occasionally looking up to reassure myself that an island hadn't appeared while I had been distracted. I listened out for the sound of waves crashing against rocks but all was silent.

After perhaps an hour, as the others slept, I became aware of a solitary light

on the horizon. Knowing no one lives on South Georgia (save the twelve people in the metropolis of Grytviken, on the northern side), I assumed it must be a ship. No sooner had I registered this than Howard Whelan's unmistakable American brogue came over the VHF calling his daughter Skye on board *Australis*. He was expedition leader aboard *Polar Pioneer*, the ship that had originally brought the *Alexandra Shackleton* to Arctowski and that would, in a month or so's time, take her back to Poland.

Unaware I was listening in, after some general father-daughter conversation Howard mentioned how amazed he was that we'd made it so far and so fast. I decided it was best to chip in at this point and let him know I was listening lest they give something away. Howard immediately greeted me warmly: "Could we come by and get a closer look at you?" he asked. I agreed. The lights approached until the previously empty horizon was occupied by a 1,000-tonne vessel and we were bathed in the ethereal yellow of its sodium lights piercing the swirling mist as several of us waved from the deck of the *Alexandra Shackleton*. It felt like an alien encounter, with this the moment immediately prior to us being experimented on by creatures with distended heads and large eyes, until I heard unmistakably human strains of "Hip hip hooray!" followed by camera flashes. Moments later, all was dark again as the lights receded into the distance, *Polar Pioneer* anchoring five or so miles away.

How different it was for Shackleton on his arrival at South Georgia. On day nine of his voyage, the painter rope was severed by a mound of ice that had formed around it and was lost to the gale, taking the sea anchor and the stability it provided with it. In addition, all six men were tormented by a raging thirst so dire their tongues were swollen and they could not touch their food. Despite their suffering, they were "irradiated" with happiness when Timothy McCarthy spied the lowering cliffs of South Georgia that, as Shackleton put it, indicated "the job was nearly done." But their optimism was premature. In view of the fact that it was near nightfall, Shackleton decided to stand off until the following morning, as, in his words, "to attempt a landing at that time would have been suicidal." Unfortunately the wind shifted in the early hours and brought with it what Shackleton described as "one of the worst hurricanes any of [them] had ever experienced." For eleven hours they battled the terrifying wrath of the storm, bailing ceaselessly. "It was the most awe-inspiring and dangerous position any of us had ever been in," recalled Worsley. "It looked as though we were doomed— past the skill of man to save." The hurricane blew them southeast until they almost ran aground on Annenkov Island. Then, when all seemed lost, miraculously the hurricane abated just in time for them to save themselves and begin tacking into King Haakon Bay from the south. Like a defiant parting blow from a defeated adversary, almost as soon as the tempest abated the pinlock that held the mast in place fell out, apparently having been on the point of doing so throughout the hurricane. In Shackleton's words, had it gone any earlier, "the mast would have snapped like a carrot" and they would certainly have been lost. "I have often marvelled at the thin line that divides success from failure," Shackleton reflected when recounting this ordeal.

For our part, we could barely control our excitement as we waited for dawn and the first sight of our promised land, but the ocean wasn't prepared to let us go either without a parting shot. While Shackleton had had a hurricane, for us dawn broke with good winds but a thick blanket of fog that meant we dared not move. Five of us were below as Seb took the post-dawn watch in the mist. Suddenly he shouted, "Land ho!," eliciting a crush of bodies through the hatch and onto deck. Sure enough, there on the horizon, still many miles distant, was the unmistakable dark outline of land coming in and out of view through the fog. It was difficult to make out what it was we were actually looking at except, of course, it was very evidently South Georgia.

What excitement and disbelief there was on board at that moment. Sun sightings and estimated positions were one thing, but evidence that we had actually made it across the Southern Ocean to this point was something else, like a lifetime of faith rewarded by incontrovertible evidence and magnificent to behold.

In typical fashion one challenge replaced another. Our latest was to work out which bit of the ninety-mile band of gray that marked the coast of South Georgia we were now looking at. Our plan to triangulate back from known reference points on the island was great in theory but totally impractical if nothing was visible more than a hundred meters above the waves. Intense discussion followed about what it was we were looking at as the cliffs loomed in and out of focus in the fog. The final consensus was that we were at the northwestern end of South Georgia. The gap between a series of headlands that could have been a bay, the far side of which we could not see, was in fact the gap between the final headland of South Georgia and Bird Island just offshore. It was important to get this right as we now headed southeast with a northwesterly wind behind us that would allow us to move offshore if we felt in danger of being pushed onto the rocks but would not allow us to come back north if we were actually farther down the coast than we thought.

Top: After the initial euphoria of seeing South Georgia, Nick focuses on the task of making landfall.

Bottom: The distinctive "Saddle Island," visible beneath the mist, indicates that Haakon Bay lies just to the southeast.

We pulled up the sea anchor and headed southeast, an appointment with destiny beckoning as we hoped that the favorable wind would take us south to King Haakon Bay. Initial progress was good as the lively wind sent us southeast, our judgment as to where we were confirmed by passing an island that, by its distinctive appearance, could only be Saddle Island.

The cliffs were dark and menacing, vaulting straight out of the ocean, waves crashing against their bases and green tussock grasses clinging improbably to any slopes less than vertical. The ocean up to a mile offshore was alive with boiling cauldrons of breaking waves, indicating multiple banks of rocks submerged just below the surface. Nick's face was racked with concern as the realization dawned that the headland marking the northern end of King Haakon Bay protruded farther into the sea than he'd thought. This combined with a marked shift in the wind to westerly from northwesterly meant that we were now in a lee shore situation that threatened to dash our tiny boat to pieces on the rocks.

Australis came through on the VHS: "You're on your own, guys; we can't go into the water you're now in without risking everything." We understood entirely and couldn't—and wouldn't—expect them to venture into the place

Boiling cauldrons of surf indicate rocks lurking just below the surface.

in which we now found ourselves, one that would be suicidal for them with their deeper draft. "I think everyone should get below with their life jackets on immediately," said Nick, his and Larso's faces the picture of focus and worry. Ed, Seb, Baz, and I did so without question, keeping our weight low so as to allow the boat to head upwind and give the helmsman an unobstructed view of the rocks all around. Sitting in the half-light below deck we spoke little, half expecting a bone-shaking impact, splitting timbers, and a deluge of green to pour into our world at any moment. It was a grim prospect, our fear being of failure as much as of the physical harm that would inevitably follow as we tried to negotiate the last hurdle approaching South Georgia.

We had all agreed back in Arctowski that the most dangerous part of this expedition was likely going to be trying to force-fit the weather we would be given with landing where Shackleton had in King Haakon Bay. Now here we were with the Southern Ocean pushing us onto this rugged, windswept coastline. All our fears were being realized in the strengthening westerly wind and the more we tried to sail out to sea to get around the peninsula, the farther out to sea it seemed to extend and the more our hitting it appeared inevitable.

Desperately threading our way through the labyrinth of breaking waves and shallow reefs, Nick aimed for the seaward end of the spine of rocks that marked the end of the headland. Both focus and faith were required as the possibility of being dashed on rocks and turned into driftwood at any moment became very real. It was a grueling literal and mental rite of passage that we were glad to have survived as, finally, after the longest one and a half hours of the voyage, we emerged into calmer waters on the northern side of King Haakon Bay, the rocks of the headland so close we could feel the spray coming off them and smell the dank vegetation. Providence had seen us through.

IMPATIENCE
CAMP

9

"To wish is little. We must long with the utmost eagerness to gain our end."

Ovid

We had endured a scare of our own as we sat in the dense fog knowing the island was so close at hand, but that was now behind us. Safely to the south of the appropriately named Krakens Teeth, a line of rocks that sit like a row of rotting molars in the middle of King Haakon Bay, a palpable sense of relief and elation began to set in. The fact that we were about to realize our dream of landing where Shackleton had was still too much for us to process; we just knew that passing the last of our serious obstacles was overwhelmingly good news. Conversations began to drift to loved ones back home and future plans as the unrelenting pressure of having made it to here afforded space to consider things beyond survival on the boat.

For the first time since we'd left Elephant Island, it appeared that nothing could now stop us. The Krakens Teeth were behind us and the risk of grounding our little boat on unseen rocks was slim. The absence of a keel, which we had all rued so many times during the ocean crossing, was now an advantage as knotted clumps of kelp slipped quietly below us, unable to get purchase on our smooth hull. Now there was no need for any mind games, hoping for the best while expecting the worst, or keeping one's guard up. Even if we grounded, we could virtually wade ashore unscathed. The *Alexandra Shackleton* had done her job, transporting her human cargo safely across the roughest ocean in the world.

King Haakon Bay was an impressive feature, ten kilometers long and several kilometers wide, with sides of vertical granite like a photograph from a geography textbook. Shackleton was similarly impressed, recalling, "The long bay had been a magnificent sight, even to eyes that had dwelt on grandeur long enough and were

Team meeting in the cave at Peggotty Bluff.

Previous pages: Arrival: towering over us are the glaciers at the head of King Haakon Bay.

hungry for the simple, familiar things of everyday life." It was reminiscent of a Norwegian fjord—appropriate since it was named after a Norwegian king—and would have made the Norwegian whalers on South Georgia feel very at home.

Approaching the head of the bay in glorious sunshine with the westerly wind at our backs, we found it difficult to remember the doubt and fear that had stalked us at every moment of the journey thus far. Now the dominant emotion was nervous anticipation of the final piece of our journey—crossing the mountainous interior of South Georgia. With our confidence buoyed by having survived the ocean crossing without serious mishap, conquering this terrain—familiar territory to me—would surely be a mere victory lap. How wrong that would prove to be.

The ice at the head of the bay was separated into two streams: the smooth ice of the Shackleton Gap and the heavily crevassed ice of the Briggs Glacier. The first was a moderately steep icy slope bookended by a rock-strewn bank at its base and a towering cloud-laced nunatak at its summit several kilometers inland; the second was a glacier so chaotic that even the desperate Shackleton and his men would not have tackled it. Anyway, this would be another day's problem. For now the beach beckoned and spirits were high.

We progressed toward shore rapidly and suddenly there was little time for the landing party to position themselves to jump in and guide the *Alexandra Shackleton* home. Elephant seals casually glanced at us, unaware that their home was about to be invaded by six humans. The fact that our fetid smell and filthy appearance were unthreatening to them gave a clue as to the state we were all in. Certainly we'd been living like them below deck for the past several weeks.

Adrenaline coursed through us, a charged mix of anticipation and apprehension. We couldn't afford to get the landing wrong: the *Alexandra Shackleton* weighed three tonnes with the combined weight of boat, crew, ballast, and equipment, and that was enough to crush a man, not to mention damage the boat against the rocks. Shackleton, knowing that it was far too dangerous to attempt to sail around to the inhabited side of the island, had dismantled the *Caird*'s topsides to make her light enough to haul up the beach. In the process of dragging her over the rocks the men had broken two masts but, according to Worsley, "did not grieve as [they] had had enough sailing to last [them] a lifetime." We, on the other hand, couldn't afford to damage the *Alexandra Shackleton*. She still had one more onward journey—to Grytviken, on the sheltered northern side of South Georgia, where she was to be collected a month hence by *Polar Pioneer*.

Top: All smiles as we head up King Haakon Bay.

Bottom: Nearly there: invading the home of the elephant seals.

The three men assigned to plunge into the icy water to guide us in would be Nick, Larso, and Seb, the sailors whose role would now be to support the climbers—me, Baz, and Ed. They jockeyed for position on the prow, looking for a suitable landing place in the clear, knee-deep water, then jumped in, immediately struggling to control the boat even in the light swell. As they steadied her, Baz, Ed, and I jumped off the *Alexandra Shackleton* onto terra firma, the shingle crunching with a reassuring solidity under our boots.

We had done it. Dazed, we hugged one another, unable yet to grasp the enormity of what had been accomplished. Then, like free-falling parachutists, we regrouped in a huddle, shutting out the view of the camera crew who had landed just ahead of us intent on encouraging "spontaneous" celebration. Our group hug perhaps represented our desire to shut out the world just for a few seconds while we tried to appreciate what we'd achieved as a team. It was Sunday, February 3, twelve days since we had left Elephant Island, although this didn't mean much to us. We had come from a time capsule with little appreciation of time in any conventional sense; it felt as though we had been on board for months.

As we broke formation, the two cameramen singled out Nick and Larso and began asking them for their thoughts on the expedition. I approved of the selection of skipper and navigator for the first interviews, as they had much to say on the navigation and performance of the boat. Questions came thick and fast on a range of topics including sun sightings, how much water there was in the boat, and initial thoughts on what we'd achieved. As Nick and Larso fielded them as best they could, Baz, Ed, and I wandered down the beach toward the Shackleton Gap, trying to size up the challenge that lay ahead and to imagine ourselves picking our way up its icy slope.

We walked drunkenly down the beach like marionettes, lifting our knees high with each step, our legs buckling with each footfall. Nearly a fortnight of cold and inaction, combined with numb feet and bodies accustomed to compensating for the roll and motion of the boat, now threw us off balance. The Gap for now looked innocuous enough, although I was surprised at the extent to which the ice had melted on its lower slopes.

We returned after twenty minutes to find the interviews still going on. Even from a distance I could see the telltale signs of cold beginning to creep into Nick and Larso's body language as they shifted their weight from one foot to the other impatiently. I approached, gesturing off-camera to my wrist to let the camera crew know the importance of winding the interviews up quickly to ensure my men

Team huddle on the beach at Peggotty Bluff.

didn't get any colder. The cameramen acknowledged me somewhat dismissively: for them this was classic television and it was important to get a complete download while emotions were raw and unsullied. I could see their point and let it go but was conscious we needed to keep on top of the distraction of filming now that we were back on dry land. I was also aware that Seb had not yet started his interview and was getting progressively colder, sitting as he was wet on the beach.

With the first two interviews over, Nick, Larso, and I decided that the three sailors should live aboard *Australis* from now until we began the land crossing. In the meantime they had got extremely cold, having stood for a considerable time in thigh-deep water as they maneuvered the *Alexandra Shackleton* to shore and then onto a towing line to take her out to be moored alongside *Australis*. They then had been subjected, still wet, to lengthy interviews. To add insult to injury, it had now started raining heavily. Dry, modern gear and a hot meal awaited them on board *Australis*—important as we needed them to remain in good physical shape. It seemed an innocuous decision and the right one at the time, but it was one I would come to question in the following days.

Meanwhile, Baz, Ed, and I had decided against erecting tents and establishing a camp on shore. We'd be soaked by the time we got our tents, gear, and food

from the hold of *Australis*, then found an area that was sheltered from the wind but far enough away from the fur and elephant seals that had commandeered all the best spots in the tussock grass. If the weather lifted overnight, we'd still be too wet to attempt the crossing. Soaking wet clothes at sea level are one thing but where "storm demons work their wild will and wreak their fury," as Worsley described the snow and ice of South Georgia's mountainous interior, they would be disastrous. Better that we stay on board the *Alexandra Shackleton* while the sailors assume their new identity on board *Australis*, which had anchored several hundred meters offshore in the lee of the Vincent Islands at the head of the bay. With Seb's interview complete, he, Larso, and Nick boarded the Zodiac and set off.

Next, Baz, Ed, and I got our interviews out of the way. Many of the questions related to how we were feeling. To be honest, I couldn't articulate my feelings because the enormity of what we'd so far achieved hadn't yet sunk in. There was, above all else, simply a sense of relief at having made it. We could stop dreading what the ocean might throw at us and knowing that our survival relied on the judgments of others, regardless of how much we trusted them. Baz and I suggested that the wet and cold were now beginning to take hold. It was our way of saying

Baz and I, unsteady on our feet after two weeks on the boat but happy to be on land.

that the interview was over as far as we were concerned. With that, we returned to the *Alexandra Shackleton*.

With only three on board, she suddenly seemed spacious compared to what we'd become accustomed to for the past few weeks. Shutting the hatch as the rain continued to fall, we fell asleep after an army ration meal and a couple of tots of Mackinlay's to the surreal sound of merriment on board *Australis* only ten meters away on the same mooring line.

The following day the heavy rain and wind continued unabated. Life below deck continued unchanged, save for the fact that being in full view of *Australis*'s living area was less conducive to hanging your backside over the edge of the boat for toilet duties than the unbroken horizon of the Southern Ocean. Still, there was no chance of setting up a tent onshore, so we used our time well, reorganizing the clutter of wet clothes, food, and unclaimed gear that lay about the boat. "Gear before beer," Baz used to say, adopting the Marines' edict of returning things to order before starting celebrations. We could be forgiven for not having done this earlier—it's not every day you cross the Southern Ocean in a replica *James Caird* using traditional navigation.

By lunchtime we had sorted and re-sorted everything to our satisfaction and were beginning to tire of being stuck on board, unable to get to shore or to *Australis*. Sleep punctuated by conversations about gear and weather was the order of the day—that and Ed's concerns about his feet, which were stubbornly refusing to come back to life. He had stoically put up with them until now, but Baz and I both knew things were not good. Trench foot in the wet, cramped, and cold conditions on board had affected everyone, but Ed was suffering more than most, perhaps because he'd decided to keep his boots on more of the time in readiness for filming at a moment's notice. Whatever the case, he'd lost sensation in most of his toes and thus far there was no sign of it returning.

Our slumber was broken by Nick's voice on the radio suggesting I come to *Australis* to discuss something. Although keen not to break the spell of living the traditional way, I realized something was amiss and that it should not be the subject of a radio conversation broadcast to all or a shouted conversation between the stern of *Australis* and the *Alexandra Shackleton*.

After the world's shortest Zodiac journey (ten meters), I boarded *Australis*, feeling immediately out of place in my wet, drab, foul-smelling clothes. Nick ushered me to his cabin and got straight to the point. "This is a very difficult and sad decision for me, but I can't do the climb." His words were typical of him—clear, measured, and unambiguous. They came as a blow but not a shock. Nick's problems with his feet had gotten worse, his toes were swelling and he felt a lumpy discomfort in the soles of both feet. The doctor confirmed he had serious trench foot and that the consequences of him exposing his feet to further cold or extended periods walking or climbing in South Georgia's icy interior could be dire. The lack of circulation could lead to the more rapid onset of frostbite, loss of toes, and even gangrene. Given the inevitability of his being exposed to conditions that would worsen the state of his feet, and the fact that he perhaps felt his work was largely complete with his having skippered us successfully to King Haakon Bay, I completely supported his decision.

I spent the next few hours deep in thought, with an increasing sense of unease borne of a combination of the continued rain, Nick's news, the feeling that control of events was somewhat out of my hands, and the inevitable low after yesterday's high. Late in the afternoon Nick's voice again came over the VHF. He sounded somber, saying he wanted to come aboard to have another face-to-face discussion, this time about the weather. Minutes later the radio crackled into life with the film crew expressing their intention to join him, at which point I instantly assumed the worst. They would hardly appear if the news were good: there was no jeopardy in good news. Alarm bells rang as I realized half of the *Alexandra Shackleton* crew and the film personnel had been reviewing Ben's satellite weather information and interpreting this in the context of the land crossing, not just for when was best to tow the *Alexandra Shackleton*. It was something I naively hadn't considered. What had seemed an obvious decision to have the sailors from the *Alexandra Shackleton* live aboard *Australis* was, I realized now, going to cause some challenges.

The satellite weather information being received by *Australis* was relevant to two aspects of the expedition from this point on. First, we needed to know when the weather would be good enough to tow the *Alexandra Shackleton* around the northwestern end of South Georgia to the more sheltered northern side from where she would later be collected. Light winds of twenty knots or less were required, especially if they were headwinds or were different from the prevailing wave direction. Second, there were the implications of what the same weather information meant for planning our land crossing. The first application I was happy with, the second somehow didn't feel right to me, as it seemed to give us an unfair advantage over Shackleton. It was impossible, however, to look at the weather information for one aspect without making judgments as to how it affected the other, although the data was actually only good for up to ten meters above sea level, making it unreliable for predicting conditions in the high interior of South Georgia.

I mulled over Nick's request for a discussion with the uncomfortable realization that I couldn't control access to the weather information on board *Australis*. I also wondered to what extent the film crew would start to impose on us now that we were accessible again. I suspected I was about to be presented with weather information I didn't want to hear and people's judgments based on that information; I wasn't happy about the prospect of either.

An hour later, it was crowded below deck, with Baz, Ed, and me having to revert to life with multiple bodies crammed together as our guests wriggled into the semidarkness through the hatch. After much commotion, they all contorted themselves into position. This group of seven sat squashed uneasily in an area about the size of a kitchen table, as if about to enter into awkward peace negotiations. There was me, Baz, and Ed plus Nick, Joe the cameraman, Dr. Alex Kumar, and Jamie, the series producer, complete with camera.

Nick kicked off first, the camera light illuminating his face in a strange white glow as he spoke. "The weather doesn't look good for the climb," he offered, Joe nodding in support as the camera focused on my face, trying to capture my thoughts about the news. Nick went on to ask me if I wanted to know what it held in store for us. This was somewhat rhetorical: I knew I was likely to be told anyway and that people's reactions would speak for themselves if I suggested we begin the climb when they were in possession of information that suggested to them we shouldn't. I told him to go ahead. Nick revealed that Ben's satellite imagery showed that the first towing opportunity for the *Alexandra Shackleton* was going to be lunchtime on Sunday, six days away. If she left at noon, *Australis* could be in position with the *Alexandra Shackleton* by 1 or 2 A.M. on Monday, February 11, at which point we could start the climb. Joe added his support to this being the best option, "recommending" that we wait until Sunday to do the crossing. My interpretation of this was that Joe had no intention of traveling before Sunday, regardless of my decision. The wind then would apparently be only twenty knots from the northwest with little rain—good for towing and theoretically good for the climb.

Everyone waited for my reaction. I was annoyed on so many levels, I didn't know where to start. I didn't like being put in a position of disadvantage by not having information in front of me on which to base a decision. Nor did I like that detailed discussions about the weather and the best time to cross

South Georgia had been taking place without me, the result being that everyone aboard *Australis* had formed his own opinion as to the best course of action, independent of Baz, Ed, and me. Since we were the ones actually doing the climb the Shackleton way, I felt it wasn't even their decision to make. Resentful, accusatory thoughts raced through my mind, jostling with a more reasoned voice that tried to give everyone the benefit of the doubt. In innocently looking at the information in order to determine when to tow the *Alexandra Shackleton*, they couldn't help but consider what the same weather might mean for the climb. As they were going to be climbing across the mountainous interior too, perhaps they would have been remiss in not having looked at the weather and then asking me for my opinion on it. Nevertheless I felt irritated at what had just happened and wasn't entirely sure why.

I didn't like that the weather information was being used to determine when we should climb as much as when to tow the boat. I was also annoyed with myself for not having set out clear guidance on who should have access to weather information and what it should be used for. Now I didn't like the fact that this information was being given to me in the form of an ultimatum, with questions about whether we should wait until Sunday or Monday being put to me in a rhetorical fashion. Most of all, I didn't like that the main, self-appointed arbiter of taste and judgment about the crossing seemed now to be the TV crew, who in reality were there to film what happened, and not to say when or whether things should happen. I hated too that the camera was rolling and that all of this was being used to get a reaction from me for the purposes of the film. That so much dissent and disharmony were being stirred up only forty-eight hours after having pulled off one of the great feats of ocean navigation and survival felt deliberate for the purposes of creating film drama, and I hated it.

Although not really the point, the reality was I knew too that the satellite weather information was only good for up to ten meters above sea level, the pressure isobars and rainfall contours for the island simply being applied as if South Georgia was more ocean rather than a massive island with its own unique weather. Plus forecasts five days out couldn't be relied on. In all likelihood we were arguing about information that was of little use for our purposes anyway.

As I tried to process all my conflicting thoughts while keeping a lid on my frustration, the doctor chimed in with some unwelcome news of his own. The camera focused on him and Ed as he calmly revealed to Ed that he was unlikely to be covered for the climb by the film company insurance due to the severity of his pre-existing injuries. On that basis the doctor's professional recommendation was that he not go. My instinctive reaction was that this whole scenario was being conjured up for some on-camera drama, but when I saw Ed's dismayed reaction it was clear that the news came as a bolt out of the blue. Again multiple thoughts crossed my mind. Had Ed been keeping the extent of his trench foot injury from me in order to keep his place on the team? If so, I could understand it—revealing the pain he was in might have forced me to take him off the team, and I knew how much doing the land crossing meant to him. In addition, if he didn't take part, what would this mean for the historical accuracy of the expedition? What would it mean for our permits and our insurance, all of which clearly stated that we would cross as a three-man team in authentic clothing supported by a team of five in modern gear? Moreover, could the doctor make these determinations, effectively ruling Ed out, or was the final decision still Ed's or mine to make? I wasn't at all sure.

I had been presented with a bewildering amount of information on which to make decisions but knew two things immediately. First, I hadn't been in possession of the facts long enough to make a clear decision on either situation and felt I had been put in this position by those who were keen to get a shocked reaction. Second, the film team was seemingly calling the shots about both when

The row of rocks in the distance are Krakens Teeth, the last obstacle to reaching King Haakon Bay.

to climb and Ed's participation. This combined with Nick's withdrawal meant that our permits and insurance were likely to be invalidated. I felt the film team had overstepped the mark and my anger erupted as I told Jamie in no uncertain terms to turn off the camera. He protested, but I wasn't going to give anyone the satisfaction of filming my reaction and reaction it was.

With an intensity that surprised me, I let rip about much of what I had heard. I felt drama and dissent were being generated that could potentially undermine the expedition, and I found it unacceptable that detailed conversations about the future of the expedition had taken place without me with decisions virtually having been made already. Not least of all, I objected strongly to the use of modern weather information to make these judgments and felt there was a distinct lack of commitment to the task at hand. When I had finished, people left quietly, there being nothing more for anyone to say.

The following morning, having decided that the Pandora's box of satellite weather information was well and truly open and my hand forced, I had a look at it myself to ensure I was better informed. I called a meeting in order to clear the air, refocus our energy, and put an end to the decision-making-by-committee approach that seemed to be taking over. Ed also gave me the good news that he was insured to do the climb, it being his decision as to whether to live with the risk of further injury, which he'd decided to do. The aims of the meeting were to remind people what we had achieved so far, what our goals now were, and to set out some rules as to how we needed to operate from this point on. I chose Travis Cave as the meeting venue, as much to keep out of the constant rain as for dramatic effect. Curiously Shackleton never mentioned the cave, despite it being an obvious place in which to shelter, avoiding the need to be bent double in a tent or, in his case, under an upturned boat. Everyone chose a rock and listened intently as the surf crashed only meters away.

I started with the weather. To me, an early morning Friday departure in forty-eight hours' time looked viable, with twenty- to thirty-knot northwesterly winds and some light rain and snow predicted over the subsequent twenty-four hours, if the weather model held true. Short of no wind and blue skies, it was as good as might reasonably be expected on a sub-Antarctic island. We'd have a wind at our backs and wouldn't be too wet. I reminded people in no uncertain terms that we were here to retrace the most difficult survival journey of all time and not to go for a Sunday afternoon stroll in the Cotswolds. I also reminded those with climbing experience who the previous night had questioned the weather

Clockwise from top left: After our trip we were too smelly even for the seals; the leather boots we'd be wearing on the walk over the glaciers; keeping warm in Travis Cave as we wait out the weather.

that we were doing something that under normal circumstances you wouldn't advise anyone to do anyway. This, therefore, required a change in mind-set. As these northwesterly winds were too strong for towing the *Alexandra Shackleton* for the first seven or eight hours of her journey to round Bird Island at the northwest end of South Georgia, she would unfortunately have to be collected later. If worst came to worst and the weather window for the climb changed, we could retreat back down the Gap to King Haakon Bay. *Australis* could be back by Saturday and thus still take advantage of the Sunday towing window. The climbers could then regroup and leave again early on Monday morning. Effectively this gave us two bites at the cherry.

In terms of injuries I wanted to be kept informed of any developments and wanted no more talk of such injuries invalidating people's insurance cover. If someone felt they honestly couldn't go on, I would understand, but it wouldn't be insurance that held them back. It would come down to an individual's personal decision as to whether he was prepared to live with the risk of what we were doing, and I reminded everyone that we all knew what we had been letting ourselves in for when we signed up on this project. And we had all been given details of the insurance; if some people hadn't taken the time to read it, that was not my problem. It was important, however, to realize we were all in this together, our permits and our insurance being based on us undertaking the expedition as a team. We had two lots of insurance: that which I had taken out to cover every aspect of the expedition and that which Raw had taken out for its people, between which there was considerable overlap. I was at pains to state how long it had taken me to negotiate the insurance and how much it had cost. It wasn't done on the back of a crisp packet; this was an exhaustive policy that everyone had been made aware of before we left. It was time to step up and do what we came here to do.

I went on to comment on one or two of the disparaging remarks I'd heard from the film team about the historical authenticity of the tented accommodations that Baz, Ed, and I were by now in on the beach and the food from *Australis* that we were consuming. I reminded them all that while Shackleton had lived under the upturned *Caird*, we could not do the same because we had to relocate the *Alexandra Shackleton* to Grytviken undamaged as part of our permit obligations. Dragging the *Alexandra Shackleton* onto the beach would obviously not allow us to do this. With regard to food, I asked for alternative suggestions as to how we could supplement our diet while at Peggotty Bluff short of clubbing elephant seals and eating albatrosses and their eggs as Shackleton had done. Save for the sound of breaking waves, there was silence, which I took as agreement.

I finished on what I hoped was a high note. We had achieved so much and should never forget that, and we would be making history if we could only keep focused for a bit longer. We had to finish this great expedition of ours and do so with the resolve we had shown across the ocean and in the years spent planning the trip. There were no men on Elephant Island relying on us making it, but we must approach our journey as if this were the case. If we didn't make it we'd be letting down not just twenty-two men, but rather ourselves and the thousands of supporters who stood behind us. I saw nods of approval and perhaps a few knowing looks on some faces as I referred to the importance of not causing issues where none need exist. When I opened the floor to any dissenters, there was silence.

That night Ed, Baz, and I polished off a whole bottle of Mackinlay's in Travis Cave, singing and telling jokes by our driftwood fire in the hope that the meeting had cleared the air and that we stood on the verge of achieving our dream. Finding myself in this situation of needing to manage factions with conflicting views made me think of Shackleton's leadership skills after the sinking of the *Endurance*. My

South Georgia's mountains and their unpredictable weather—never to be underestimated.

Tough decision:
with his feet in
rotten shape, Ed
made the call
not to do the
mountain climb.

admiration for him soared. Just as he had to guard against the breaking up of the ice floes on which they camped, so he was constantly on the lookout for cracks in the unity of his men, cracks that inevitably began to show with the strain of their ordeal. Reginald James, the expedition's physician, noted that Shackleton "was constantly on the watch for any break in morale or any discontent, so that he could deal with it at once." His care of his men and determination to keep spirits up were indefatigable, and his ability to maintain his optimism in the face of seemingly insurmountable odds now struck me more forcefully than ever. By comparison I was faced with relatively minor discontent and doubt among my team.

I mused on how Shackleton managed dissent and stress among his men. He regularly served hot, sugary milk and food to soothe away troubles, and gave the best of their limited gear to those who needed it most. But he also read the riot act to all the men after McNeish questioned his authority on the ice. Perhaps the secret of his success was a combination of his keeping "a mental finger on each man's pulse," in Worsley's words, and his utter conviction whenever he decided on a course of action. Perhaps one could call it stubbornness. Orde-Lees commented, "It is no use trying to convince Sir Ernest if he had formed an opinion of his own." Whatever the case, the strain began to show on Shackleton. Worsley wrote of their time waiting at Peggotty Camp: "Meantime, the strain of waiting and anxiety for his men was telling on him. He was then more discouraged, worried, and nearer to depression than I had ever known him. He said to me one day: 'I will never take another expedition, Skipper.'" I knew how he felt but, like Shackleton, was damned if I would give up now. Whisky seemed an appropriate anesthetic for my frustrations and I was sure the Boss would be raising a glass in spirit with us.

Late the following morning, we moved our tent closer to the Gap, as activity on *Australis*'s deck indicated she was getting ready to begin her twelve-hour journey around to Possession Bay. Soon our tent was joined by two more, one containing Si and Joe, the other Larso and Seb. Our aim was to depart at 10 P.M. so as to give us two hours of light in which to cross the rocks that led to the ice at the base of the Gap. We felt a mixture of relief and excitement to be getting ready to leave. Ed, in the meantime, had been aboard *Australis* and now walked toward us along the beach looking hangdog, with Jamie close behind, camera in hand. "Here it comes," whispered Baz under his breath. Sure enough, Ed wearily reprised Nick's words about needing to pull out: his feet were too painful to walk on, regardless of him being covered by insurance. Ed was tough and uncomplaining, but the redness in his eyes indicated how disappointed he was. I consoled him as best I could, not really aware of what this might mean for him and the expedition, not to mention the film. His expedition now over, he embraced Baz and me before hastily returning to the Zodiac, almost wanting to distance himself from the scene of his disappointment as quickly as possible. Shoulders hunched, receding into the distance, he cut a dejected figure, as if the Zodiac were transporting him to a prison hulk.

The fur seals nipped disrespectfully at our ankles, unaware of the enormity of the decisions going on in our little world. I wondered who had the high ground here: we humans, who with our superior intellect were voluntarily attempting this craziest of trips, or them, who made this harshest of places their home. It didn't take much to imagine their forebears having prophesied the return of these inappropriately clad humans.

CHARTING THE COURSE

Celestial navigation involves working out where you are based on observation of the positions of the sun, moon, planets, or stars. You do this by knowing which point on the rotating earth a celestial object is above and measuring its height above your horizon. To do this using the technology available to Shackleton in 1916, you require a nautical almanac that lists the position of celestial objects at given times, a marine chronometer that keeps accurate time while at sea (the clock is on a "gimbal," which prevents the rock of the boat from making the pendulum in the clock speed up or slow down and affect the chronometer's accuracy), and a sextant to measure the body's angular height above the horizon.

That height can be used to compute distances from the subpoint, to create a circular line of position. A navigator shoots a number of celestial objects in succession—either stars, the moon, or in our case the sun to produce a series of overlapping lines of position. The point where these lines intersect is known as the celestial fix.

In order to measure longitude (how far west or east you are), the precise time of a sextant sighting must be recorded. Each second of error is equivalent to fifteen seconds of longitude error, which at the equator is a position error of a quarter of a nautical mile, which is about the accuracy limit of manual celestial navigation.

Latitude (how far north or south you are) is easier, particularly if you can measure the angle to the sun at your local noon when the sun is at its highest point, which is what we did on the expedition.

CALCULATING LATITUDE: If you imagine the sun is at the top of a giant pole and the height of the pole and the location of its base are known (provided by the almanac), you can measure the angle between where you are and the sun and calculate how far you are from the base of the pole. This will give you your latitude. A second bearing on the sun can be taken and this will theoretically intersect the first latitude fix you took and give you your location.

The one time during the day when you can obtain a precise bearing on the sun is at local noon, when the sun is as high in the sky as it will get that day. At that moment the sun is either directly north or south of you and any position fix from it will coincide exactly with your latitude. Finding latitude from a noon sight is therefore the easiest celestial navigation calculation and provides very accurate results. It works as follows:

1. Subtract the sun's altitude at your local noon from ninety degrees (the total number of degrees of latitude in your hemisphere). This gives you what is called your Zenith Distance (ZD).

2. Look in your nautical almanac to see what the declination of the sun (or "latitude of the base of the pole") is to the nearest hour of when noon occurred at your location.

3. Add this to the ZD if you and the sun are in the same hemisphere; or subtract it if you are not. This result is your latitude.

CALCULATING LONGITUDE: Obviously, if the sun is directly north or south of you at noon, you and the sun must be on the same line of longitude at that time. If you know exactly what time noon occurred at your location, you can simply look in the nautical almanac to find your longitude. This will be the same as the sun's Greenwich Hour Angle (GHA), which is the equivalent to the longitude of the base of the sun's pole.

The tricky part is getting the exact time of the event. The sun moves quickly (or rather the earth turns quickly), and even a couple of seconds of timing error will throw your position fix off by several miles. Even if you have a perfectly accurate watch, getting an exact time for noon is problematic. The sun always seems to hang at its highest (or meridian) altitude each day for a couple of minutes or so, and to time noon precisely you need to figure out exactly when the middle of that "hang time" is.

The technique to do this is fairly straightforward. I get out the sextant fifteen to twenty minutes before I expect noon to take place (I can easily get a rough estimate from the almanac) and take a leisurely series of sights as the sun climbs higher in the sky. The sights I feel particularly good about, I mark with an asterisk. After the sun reaches its meridian altitude, I preset the sextant to the altitudes for the three or four sights I felt were my best when the sun was going up and carefully record the time when the sun again hits those altitudes going back down. I then calculate the time spreads between each set of matched sights and figure out the time of the midpoint of each spread. These times I average together to arrive at a best estimate of when precisely noon took place. There are then various corrections you need to make to make any raw sextant sight accurate. These account for things like which part of the disc of the sun you've measured and inaccuracy in whatever timepiece you're using (in our case a chronometer that did lose and gain time), et cetera.

PAUL'S TAKE: A lot of the theory of a full course in astronavigation could be disregarded for the purpose of navigating the *Alexandra Shackleton* from Elephant Island to South Georgia. Obviously we read every account we could of the journey, paying close attention to Worsley's own. It became obvious that the theory behind Worsley's plan was directly applicable to our own journey. The logic was as sound then as it is now, so given the same tools, we followed it. The fact is that there were certain pressures on the navigation that made some of the choices obvious, and I'm sure we would have arrived at the same plan even independent of earlier accounts, e.g., the need to focus on getting north as soon as possible so as to distance ourselves from the pack ice . . . and ice in general.

There are many reasons when undertaking this journey to get north early. Among other things, it gives you the best chance of "hitting" South Georgia, as the average wind is most likely to come from the west. I emphasize the word "average." When missing South Georgia is not an option, you want to be setting yourself up with the best chance possible. This means coming in from the west. If, for example, you came in from the southwest and got a bout of strong NW-NNW winds, you could easily find yourself slipping below South Georgia toward the pack ice with a very real chance of missing South Georgia altogether. Remember that the *Alexandra Shackleton* has no keel and can't sail back upwind. She can make ninety degrees to the wind at best. This plan to come in from the west ties in very nicely with the astronavigation. It is much easier to find latitude using a sextant than longitude. The most basic sight you can take with a sextant is called a "noon sight." It is often referred to as the cornerstone of astronavigation. One of the big advantages of a noon sight is that it doesn't require

an accurate timepiece . . . or even one at all. While a noon sight does only give you latitude, that was perfect for this journey—the plan being to try to get as far north as possible until we were on the desired (and well known) latitude of South Georgia . . . and then hang a right turn and ride in with the prevailing winds on that latitude until we saw land.

During that time when the sun seems to hang, on a slow boat like the *James Caird* or *Alexandra Shackleton*, the latitude should not change very much, or at all if you are heading due west or east. At any other time of the day apart from noon, you need to have an accurate idea of time in order to work out latitude and certainly longitude. To be honest, I was never totally confident in our timepieces. Whereas Worsley would have been totally familiar with the ones at his disposal, we weren't. I'm sure Worsley would have done a better job of roughing it with his own timepieces if he'd had to (if no noon sights were usable due to weather, say) than we would have. That said, we did use our own clock in order to have a shot at longitude as we approached South Georgia . . . and it wasn't too bad.

Based on the aforementioned plan to head north first, we really didn't need to take any sights at all for a while. What we needed to do was use whatever wind we had to just try to climb north. We had to get a good few hundred miles, and the *Alexandra Shackleton* doesn't eat up distance that quickly. Seventy or eighty miles is considered a good day. I think around ninety-six was our best. We had a good break in the weather on day four and managed to get a great noon sight. It boosted our confidence in our dead-reckoning skills. I didn't manage to get another noon sight until day nine or ten (it's in the log), due to cloudy or impossible conditions. Some of the efforts turned to pure comedy, as everything from the sea state and cloud cover to the boat itself and helmsman conspired to throw me and the sextant overboard. It was often hard enough just to hang on to the mast, let alone take a sight. I always had one eye out for the sun in order to get a sight if possible, especially as we approached South Georgia. The fact was, there simply weren't many opportunities.

Our last noon sight did confirm that we were on the right latitude and if we stayed on it, we would hit South Georgia right by King Haakon Sound. The trouble was, with no accurate idea of longitude, we didn't know exactly when we would hit it. We had an estimate, but this vagueness becomes very concerning when you approach a very dangerous shore in foggy conditions with night coming.

THIRD-MAN
FACTOR

10

"It is our choices, Harry, that show what we truly are, far more than our abilities."

J. K. Rowling, *Harry Potter and the Chamber of Secrets*

Nervous anticipation built as the light began to fade. We would now travel as three pairs, roping up once we had navigated the initial rock-strewn landscape beneath the ice snout of the Gap and gotten higher onto the ice itself. The goal was to set off late at night and try to make it to Stromness in twenty-four hours in a bid to avoid the need to spend two nights out in the wild without a tent. If we fell short, at least the last bit of the climb, beyond Breakwind Ridge, was the most straightforward. If we could get there by nightfall, at a push we could stumble into Stromness over what was relatively straightforward ground.

Final gear checks completed, we were ready to go, Baz and I clad in old clothes, the others in modern gear. Our packs looked small and insignificant on our backs, belying the fact that they were full of the smallest and heaviest items of film equipment in an effort to ease the camera crew's load. In a nod to Shackleton's own insufficient footwear (light mountain boots into the soles of which McNeish had fitted several screws taken from the *James Caird* to give him some grip on the ice), I strengthened my well-worn leather boots, old friends from my Mawson expedition, with a number of old hobnails while waiting at King Haakon Bay. By the time he made the final descent into Stromness, the two-inch screws in Shackleton's boots were worn down flush with the soles, offering no traction and almost leading to disaster. I hoped mine would fare better.

At 11 P.M. we set off, picking our way up through the steep rocks rust red in the late evening sun. We followed the line of a watercourse that at least seemed to contain fewer boulders and provided a gravelly surface on which to walk. Baz and I were easy to follow, with sparks flying off the hobnails and screws that protruded

Baz: "It's now or never. Let's make it now." Preparing to walk the interior.

Previous pages: The Tridents await us behind the Briggs Glacier and Murray Snowfield.

from the bases of our boots. It amazed me to think that all of this rockscape would have been covered with ice perhaps 100 meters thick in Shackleton's day, all melted away by the steadily increasing temperatures of the twentieth century. Worsley wrote of "huge 'snouts' of ice projecting over 200 yards off the beach" and "ice masses of 50 to 100 tons" littering the beach under them. Certainly there was no ice either on the beach or hanging over it now. The rocks underfoot as we made our way uphill presented us with multiple opportunities for twisting ankles or worse—something Shackleton wouldn't have had to contend with, but a real risk for us after such a sedentary existence on the ocean.

After twenty minutes we stopped to adjust clothing and packs. Overheating with the steepness of the climb, I removed the heavy, woolen grandpa long johns that had burdened me for so long, and which I'd been forced to endure due to the cold conditions on board the *Alexandra Shackleton*. An hour or more of steady ascent later, we approached the imposing wall of turquoise ice that marked the snout of the ice of Shackleton Gap. We stopped to marvel at its scale, tens of meters high. Seb, Paul, and the cameramen donned modern crampons. The rear man in each pair put a small glow stick on the back of his pack to guide the others, another modern advantage forbidden to Baz and me.

The reflective surface of the ice compensated for the steadily dwindling light and, although there were cracks and holes in the ice, at least there were now no boulders to contend with. We regrouped 600 or 700 meters up the steep ice surface. Si was visibly limping and grimacing with intense sciatic pain that extended down into his back. After munching down some painkillers, he agreed to push on to the top of the pass as light snow began to fall. We had made a good start, but I could not help feeling that further problems were about to surface.

The five of us regrouped in darkness with the wind at the top of the Gap steadily increasing. The solid surface of ice had now become energy-sapping wet snow as sleet fell heavily. If we continued less than four kilometers straight down from here we would be at the sea on the northern side of the island in Possession Bay—the narrowest point of South Georgia. When Si caught up with the group it was clear he was not having a good time of it; he threw his pack to the ground in fatigued disgust.

The snow underfoot and the faintly visible outline of the big nunatak that marked the top of the pass meant that we were approaching the next phase of the climb—the Murray Snowfield that lay between the top of the Gap and the Trident Mountains to the east. The Murray, as the name implied, was a snowfield rather

The journey ahead: Shackleton Gap on the left is much steeper than it looks.

than a glacier or fully fledged ice cap, but as it might contain crevasses, we began roping up in pairs, Baz and I doing so well away from the others to ensure we got no benefit from the light of their head torches.

Si's lower back was seizing up with pain, making his right foot, ankle, and shin numb from what appeared to be intense sciatica—certainly not to be taken lightly given how Shackleton too suffered terribly from sciatica that at times completely incapacitated him. Si stretched and had something to eat and drink while roping up in an attempt to steel himself against the biting wind that was already stripping away our body heat after less than ten minutes of inaction. The painkillers apparently hadn't worked or, if they had, the pain etched on Si's face suggested otherwise.

We moved off in our three pairs into the deep snow of the Murray with Baz and me leading as usual, creating a deep, plowed furrow that was easy to follow. Instantly laboring with the workload, I found myself overheating after twenty minutes despite the cold. During polar travel it is critical to remove items of clothing before you start sweating, so that sweat doesn't subsequently freeze on you. But with Baz and I roped together as we were, this wasn't possible: not only would it have required the other man to stop, but the leg and shoulder straps of the harness Baz had made from our old rope prevented the removal of any upper layers. Against all my instincts from years of polar travel, I resorted to removing my beanie and thick woolen mitts to shed some heat.

Si and Joe had dropped back. As Joe was one of the fittest of all of us, it was obvious that Si's injury was slowing them down. We stopped to regroup, Si and Joe pulling level after a few minutes. Si immediately offered up what had obviously

been on his mind for a while: "My leg and back are too painful and I'm slowing you all down. I'm sorry, but I have to pull out." Under normal circumstances it would have been devastating news, but I had become so accustomed to difficulties of this magnitude that Baz and I took it on board and focused on what we could do. Trying to think positively, it was probably best it had happened sooner rather than later and at least we had escaped Peggotty and were under way. Si was right about the slow pace we were being forced to keep, and Possession Bay was the only evacuation point until we reached Fortuna Bay on the far side of the glaciated terrain of the Crean and Fortuna glaciers.

There was no point disputing it. Si couldn't go on. He was no novice when it came to climbing so I knew he must be in considerable pain, although the situation confirmed Baz's and my suspicions about how the film crew had underestimated the seriousness of the crossing and were not well enough prepared. Regardless of how frustrated we felt, we now had to focus on getting Si safely back to *Australis* and us back on track as quickly as possible.

Baz switched on his head torch to get an accurate reading of the prismatic compass to take us down to Possession Bay, exasperatingly more or less back in the direction from which we'd come. We turned and trudged back, taking one reluctant step after another, trying not to think about things too much. The sleet was intensifying and, as we got back onto Shackleton Gap, the wind began to strengthen. We got out the VHF from Larso's rucksack and called in the fact that

The bottom of the Shackleton Gap in the evening sun on the day we arrived, when tiredness prevented us from climbing it. The weather would never be this good again.

Si was being evacuated with a nontrauma injury only to discover that *Australis* wasn't yet at the head of the bay, having had to stop some two hours short in Prince Olav Harbor due to the danger of traveling in darkness. She would leave at dawn to come and pick up Si but not before. Baz and I conferred, shouting to each other in order to be heard above the howling wind. We agreed there was no point in us all going down to the coast if the boat wasn't there, especially as it could be done more safely in the light the following morning by just Si and Joe, saving the rest of us the need to climb down and back up again. Plus, if Baz was half as wet as I was from the deep snow underfoot and the sleet we'd been trudging through for the past hour, he must be getting pretty cold by now. He was. We agreed we were the wettest we'd been since the first big wave had broken over us at sea. We would have to stay where we were for the night.

We set up the two tents we had with us as fast as we could in the wind, battling with the billowing fabric, then dived inside and tried to get warm. Baz, Larso, and I were in one and Seb and the two cameramen in the other. Positives were hard to find. We had got going in what Baz and I had decided was decent enough weather but now, after four hard-won hours, we were hunkered down in our tents having made barely two kilometers in the direction of Stromness. The weather was deteriorating fast and we already had our third casualty, the broader implications of which I hadn't yet had the chance to digest. Clad as we were in our gabardine outers, Baz and I were wet through. Just as Shackleton, Worsley, and Crean had sunk to above their ankles in slush on their initial ascent, so our feet were saturated from trudging through the deep, wet snow of the saddle—ground we had only traveled over because of the need to evacuate Si. The far side of Shackleton Gap was, as anticipated, proving a terrible spot in which to camp—windy and exposed with strong katabatic gusts peppering the tent with machine gun bursts of hail and sleet and already threatening to collapse the tent, despite us having it pegged down with everything we had. The token warmth afforded by the single sleeping bag draped unzipped over the three of us served only to highlight what warmth our bodies might enjoy if we had the right gear with us, which we didn't.

The flaccid piece of fabric that was Paul's punctured inflatable mattress was all that separated us and the ice below. Disappointingly, it was even less effective insulating against the cold than my lowest expectations, and that was if you were lucky enough to get your third of it. I looked at my watch. Four and a half hours to go until dawn.

Familiar was the sensation of wet wool against our skin, like a penitent's hair shirt, made bearable only through the well-honed skill of lying completely still, any movement to restore sensation to the hips bringing the wool into unpleasant contact with the skin and sending shivers down the spine. Baz and I smelled like wet dogs, with Larso slightly less malodorous given his recent sojourn aboard *Australis* and the fact that his clothing was waterproof, sealing in any smells. I began to shiver hard and realized that, despite our reprising the spooning technique perfected aboard the *Alexandra Shackleton*, temperatures here at altitude as we lay on snow and ice were much colder than those we had experienced on board the boat, with the result that I just couldn't get warm.

I lay awake, taking some solace from the fact that we had made absolutely the right decision to stop where we did. Going any farther in these conditions would have sent Baz and me on a decline into hypothermia, while descending to a lower and more sheltered location would only have meant the need for us to reclaim the lost altitude when we resumed the climb at some stage. Moreover, *Australis* wasn't yet in position, rendering a hasty descent not only ill advised but pointless.

Positives could also be taken from the fact that the latest mishap meant that we had followed Shackleton's footsteps almost identically, having ended up near Possession Bay, albeit for completely different reasons—he through navigational error and we due to Si's injury. Having reached the top of the saddle, the three men had spotted through the clearing mist what they believed to be "a great frozen lake, shining in the moon's rays below." Thinking that it would afford a less

Paul securing our tent before descending to Australis in worsening weather.

challenging route than the hazardous higher ground, Shackleton decided to make for it, only to find that they were descending over the heavily crevassed surface of a glacier. Worse still, when they reached the bottom they discovered that it was not a lake at all but Possession Bay, an inlet from the sea. This mistake, which should have been avoidable given that Possession Bay was on their chart, cost them dear— there was nothing for it but to climb back up the glacier, retracing their steps for the most part and draining their energy reserves. The uncanny parallel between his experience and ours was an unexpected twist of fate: our modus operandi was to cross South Georgia following his route but not to contrive to repeat every navigational error he had made, yet here we now were, probably within a few hundred meters of where he had been. The synchronicity was striking: what we were experiencing brought us closer to what the Boss had gone through. This wasn't the first time and almost certainly would not be the last that I felt his influence on our expedition.

The weather had worsened but a wan light on the tent indicated dawn had broken. Our optimistic expectations that dawn would bring with it a lift in the weather proved to be unrealistic. The strong winds remained undiminished, although visibility had improved, revealing a rock band extending all the way down to the bay. A shout across to the tent containing Si, Joe, and Seb brought an immediate shout back, only just audible over the roar of the wind. If conditions in their tent were anything like they were in ours, it was not altogether surprising they were awake. After a discussion among the three of us in our tent, Larso braved the five meters to theirs to convey what the next steps were to be. As far as Baz was concerned, Si needed to descend to *Australis* roped to Joe due to crevasse risk and poor visibility—perhaps only 100 meters. We also agreed there was little point in Larso and Seb remaining up here in the prevailing conditions either: we needed them to be in the best shape possible in order to provide Baz and me with backup during the crossing. Larso returned with news that Si's condition had not improved and that they were ready to descend, *Australis* having begun to motor toward the base of the pass in anticipation of picking up and treating Si. It was good news under the circumstances. If we could just cauterize this situation we could take stock of our position and refocus on getting some momentum going again when the weather allowed it.

Agreeing on the next scheduled radio contact time, the band of four bundled their two sleeping bags through the door of our tent as they passed, shouting wishes of good luck as they went. Within minutes the dejected little party had

Joe and Si look for a way down the steep descent off Shackleton Gap. Australis is a dot in the distance in the middle of the bay.

been swallowed by the mist and low clouds, leaving Baz and me alone on the mountain. We zipped up the fly and clambered into their bags, lying back to consider the events of the past eight hours. The silence was broken when Baz broke into a forced rendition of Bill Withers's "Just the Two of Us." I chuckled, responding with my own version of "Alone Again (Naturally)." At least we hadn't lost our sense of humor, although our circumstances were beyond a joke.

A couple of hours later, the radio crackled into life at the agreed time with Joe voicing his concerns about his continued participation in the crossing. His reasons included reservations about being roped to two inexperienced climbers, Seb and Larso, as well as questions about the wisdom of our traveling in the current conditions. We explained we were still hunkered down in the tent, just as they had left us, precisely because of the conditions and that even we wouldn't travel in them as they were. I think Joe also had concerns about being solely responsible for recording events as the only cameraman, compounded by the weather and the glaciated terrain that lay ahead. Baz finished the call by telling Joe it was his decision to make as to whether or not he felt he could continue.

The radio safely switched off, I voiced my surprise at Joe's concerns to Baz, especially given that Joe was, in addition to Raw's cameraman, a member of the mountain rescue in Scotland. Baz casually murmured a question as he rolled over to get some sleep: "How many glaciers are there in Scotland?" "None," I replied. "Exactly," said Baz. As with previous decisions, it was important not to allow it to deflect us from the task at hand. Joe's decision was to be respected and, to be honest, we weren't that unhappy to hear it. Filming slowed us down anyway and that was likely to be a major hassle for us on the crossing. We would put up with it if we had to—it was part of the deal, after all—but if the cameramen felt they couldn't continue, then frankly it was one less thing for us to worry about. We would film ourselves with the mini high-definition cameras we'd been given. The results might be a bit *Blair Witch*, but the TV crew kept saying they wanted realism. Now they'd get it.

The wind howled all around us, stretching the guy lines as taut as piano strings as I lay with wet tent fabric pressed to my face on my collapsed windward side of the tent. Baz lay still beside me in his sleeping bag, dealing with our predicament in his own way. Neither of us spoke, but we both knew that the dream of the Shackleton double was now further away than at any time prior on this expedition. Realistically the dream would last about as long as the tent's ability to withstand the onslaught of the intense katabatic winds pouring down on us on Shackleton

Gap. It was a dark moment and I chastised myself for having lowered my guard and allowed premature thoughts of successfully achieving the double to creep into my mind. Like most, I had assumed the sea crossing would present us with all of our challenges and that, once landfall was made, the island crossing would be relatively straightforward. I was wrong. Fate had been tempted. We were in a bad place.

Anyone with a modicum of outdoor experience would know that a saddle or pass between two mountains is more or less guaranteed to be windy and unpleasant and therefore a bad choice for a camp. We had, however, ended up here through circumstance, not wishing to descend all the way to the northeast coast in darkness and heavy rain and lose the hard-won altitude we had gained. Plus Si had been unable to continue anyway, and *Australis* was anchored in Prince Olav Harbor, an hour away from the base of Shackleton Gap, restricted to travel during daylight hours around South Georgia's rugged coastline. There was nothing to do but sit it out until first light. A bit of Shackleton philosophy wasn't out of place here—the ever-expansive Orde-Lees wrote of the Boss: "He often says of a thing, 'It's time enough to do it when you've got to, until that time comes make yourself as comfortable as circumstances permit, when it does come, do it with all your will and even then make yourself as comfortable as your circumstances will then permit; comfort is only a matter of comparison after all.'"

We caught up on sleep, our clothes slowly drying out with our body heat in

Trying to be patient and remain positive while waiting out the lashing storm. And I'd thought the sea would be our biggest challenge.

the sleeping bags, waking only to put on the stove to make up some hot milk as the wind seemed intensified despite us willing it to abate. The tent fabric on the windward side was now pressed onto my face at an improbable angle, requiring me to lean toward Baz just to be able to breathe freely. At some point during that timeless period in the tent the radio came to life again. This time it was Seb, and I could tell by the tone that we were about to revisit territory we'd been through with Nick, Ed, Si, and Joe. "Boss, this is one of the most difficult decisions I've ever had to make . . ." Seb, too, had succumbed to trench foot and, like the others, was concerned that it could get worse and lead to frostbite if he continued. As I had with the others, I respected his decision. There was nothing to do but be patient and weather the storm.

Things were starting to get very serious. I looked at Baz and wondered what it meant for us continuing. I waited until 6 A.M. Australia time and called Kim on the satellite phone, asking her to find out if we'd still be covered by insurance should we continue unsupported. Kim confirmed that the insurance company had, to our surprise, conceded that we would be, even if Baz and I went on just as a two-man team in old gear. It was a sobering prospect but we were ready to do it.

Several hours later, Larso radioed through. The message had got back to him through e-mail communication between Kim and our blogger Jo that Baz and I were going it alone. Ironically, on an expedition using hundred-year-old

technology with team members less than a mile apart, he had got the news from a web feed. I detected some frustration in his voice—understandable if he felt a decision about his fate had been made without his input. In the tones of a diplomat attempting to smooth over a brewing international incident, I reassured him that that was far from the case; it was just crossed wires. We had merely been asking whether we were able to go it alone because it had looked as if we might have to, but our preference remained emphatically to do it with him. We had assumed that he was understandably considering his options, given the atrocious weather conditions and the fact he was a sailor rather than a climber. One by one, the others had dropped away; we had to consider that he might have felt the same way. The edge to his voice softened a little as, appeased, Paul told us he had absolutely kept his "race face" on and shouldn't be ruled out by any stretch. This was fantastic news. We agreed we'd speak again in a couple of hours with a weather update for him and any further thoughts he had from his point of view.

There was a sense of loneliness up there in the mountains but somehow it was cleaner and easier to deal with now that we were just two—the strongest pairing when it came to traveling through such terrain, with both of us on the same page in terms of our desire to do this thing and neither stealing any of the other's precious motivation in order to maintain focus. In addition, both of us were comfortable in the extreme weather we were engulfed by despite it being even worse than either of us had anticipated, the wind now gusting at over 155 kilometers an hour. Despite our dire circumstances in these savage conditions, I suddenly realized that it felt as if a burden had been lifted from my shoulders now that managing problems associated with a larger team of people had, for the moment at least, evaporated. I had rapidly come to realize that numbers were our enemy. As Baz had said about taking large numbers of soldiers out into the field, if you take enough of them, someone will always have a problem; it's just a numbers game.

In the meantime, the weather was truly awful. Lulls in the wind in turn lulled us into a false sense of security that things were improving, faint hopes that were dashed as soon as the next gust of wind poured down through the rocks at the top of the pass, slamming into our tent seconds later. It was like being back at sea, such was the force of the wind that hit us like waves of water.

Things were about to take a turn for the worse as the intensity and frequency of the gusts increased. Suddenly and without warning following a lull, we heard the next gust approaching, an enormous pressure wave hitting the side of the tent, savagely tearing the tent poles from the fabric and collapsing it onto my face. The

wind, previously deflected by the slope of the windward side of the tent, now found its way under the groundsheet, lifting our temporary home and rolling me inward toward Baz as the whole structure collapsed. In a matter of moments, we had gone from relative calm inside the tent to being exposed to the raging elements, wrestling a billowing, formless piece of fabric with wind screaming all around us. Extricating ourselves from our sleeping bags with extreme difficulty, like parachutists under a collapsed canopy, we fumbled for the unseen door, managing finally to unzip it and launch ourselves horizontally feet-first into the maelstrom, donning our boots as we went past. Baz had been wearing down tent boots that were part of the emergency kit in the tent, but one of these was immediately torn from his foot by the wind as he exited, disappearing downwind at enormous speed and leaving his socked right foot exposed to the elements as we surveyed the scene of destruction wearing only our sweaters. The surface of snow was alive with snow streaming horizontally down toward the coast, our tent torn beyond repair and lying flattened and formless without our bodies to give it structure. These surely were Worsley's "storm demons [that] work their wild will and wreak their fury."

It was tumult all around, the nunataks at the head of the pass looming in and out of view as clouds poured over them. Our tent was finished but the cameramen's tent was still intact only meters away—a not insignificant distance as I set about rummaging for our gear in the remnants of our sanctuary, sandwiched between the roof and floor that were now as one. The loss of his boot had forced Baz to seek cover in the cameramen's tent and try to get sensation back in his toes as I threw in recovered items for him to organize. Ten minutes later I dived in, brushing off the worst of the centimeter-thick ice that had adhered to my sweater and zipping up the door hurriedly. We hadn't abandoned our tent as much as it had abandoned us, but at least relative calm again prevailed in our new abode. For now.

We sat there exhausted and as near as we had been to truly despondent, Baz getting a brew of hot, sugary milk going to warm ourselves up. As we tried to gather our thoughts, the radio again crackled into life, an incongruous intrusion from another world and another era ironically no more than a mile away. Baz answered the call. It was the doctor, wishing to let me know that, as I had pre-existing frostbite damage to my right foot from a prior expedition, I should undergo an assessment to determine my fitness for the task ahead. Any feelings of stunned outrage I may have felt at this were tempered by a sense of real contempt. It was clearly an unwelcome intervention but not one I felt pressured by in any way: there was no way the doctor would be able to get up here in the current conditions, I didn't want to see him, nor, more to the point, would such an assessment serve any useful purpose. I knew I had previous frostbite damage and it made precisely no difference to me. Moreover, the doctor's stilted tone indicated that cameras were rolling at his end. As a result, I treated the call as one he felt he needed to make in order to protect himself against liability, and perhaps also to create a healthy dose of jeopardy for the benefit of the cameras, neither of which struck me as very admirable reasons. My contempt at what seemed to me the predictable transparency of the call hardened as I told him in no uncertain terms that I didn't need this nonsense, topped and tailed with a few Anglo-Saxon expletives for good measure. Realizing I was not to be moved, he read out what sounded like a prepared statement: "I just want to confirm that you have declined my medical assessment." I replied that I had, saying nothing more.

Again I assessed where we were at. On the downside, we had lost twenty-four hours of time and counting, as well as 250 meters of hard-won altitude and, despite all of our efforts, were no closer to Stromness than when we started. Now we had lost our tent, and the majority of our team, including our cameramen, and this abject weather that had us pinned down so mercilessly, leaving us soaking wet and bitterly cold, didn't seem to be abating. Forces that Shackleton hadn't had to cope with were also threatening the expedition, principally in the form of injury-induced abandonments and the implications that these might still have for the integrity of our permits and insurance cover, implications that our doctor seemed only too happy to tell us about for what I regarded as being the wrong reasons. Another issue that had the potential to bring the expedition to a grinding halt was the fact that our charter of *Australis* ran out in eleven days' time, at which point we needed to be in Port Stanley, yet we still had the climb and the return journey from Stromness to King Haakon Bay and then back to Grytviken with the *Alexandra Shackleton* in tow all to do, followed by time sorting things out at Grytviken and finally a four- or five-day crossing to the Falklands. Things were getting tight.

On the plus side, we were a lot more comfortable now that we had acquired Si and Joe's sleeping bags, replaced our punctured Therm-a-Rest, and were in a tent that might actually survive the brutal conditions that had destroyed the other one. Plus, we were hoping, not unreasonably, that wind speeds could not go much higher. Most of all, we were still here and the more the odds stacked against us, the more our resolve to complete the challenge strengthened. This combined with the detour down to Possession Bay made me feel I was getting closer to Shackleton all the time, which I liked. He, of course, would have viewed these problems as "just things to be overcome," to use his own words from his Nimrod diary, and that was my philosophy too. Shackleton was, of course, compelled to continue, acutely conscious that twenty-two men were relying on him on Elephant Island. We, on the other hand, had the unfortunate "luxury" of knowing in the back of our minds that completing the expedition was in some way optional. But it wasn't. We needed to view things as if twenty-two men's survival relied on us. The reality was we had scores of people, friends, supporters, and sponsors who had placed their faith in us to do this, even if their lives were not in danger, and I wasn't prepared to fail them.

In short, the worse things became for us, the more I convinced myself that this was the way it should be. Shackleton had experienced huge obstacles, and our getting closer to what he experienced was what I wanted. If we succeeded in reaching our goal against these odds, our expedition would be worthy of people's respect and would satisfy my own self-critical view of my achievements. Not least of all, the Boss would have approved of how we were handling things.

Actually, Baz and I just decided we wouldn't let the dream go. I remembered my attempt on the North Pole and the devastating disappointment I felt at having to turn back and how it felt to have to live for so long with that defining moment when the decision had to be made whether or not to go on. I remembered, too, climbing in the Andes a couple of years before when one of my two partners had succumbed to altitude sickness and how, having left him to recuperate in the tent, the third climber and I had continued on and been rewarded by an exhilarating summit day. It was clear to me that there was no option but to keep going.

Nevertheless, we were being subjected to atrocious weather and medical problems that had claimed five of our team, not to mention mind games about my fitness to continue, and I hated the intrusion of the doubts of others into our world. Almost as a distraction I sounded out Baz as to how long he felt the dangerous conditions could be endured, any normal person being likely to respond in terms of minutes or hours given the wind, subzero temperatures, and state of our tent. After a moment he gave a measured response: "A couple of weeks." It made me laugh out loud. I was sharing a tent with an English Tom Crean. I felt the same way. There was no way we were not going to do this thing.

The next radio call was with Larso a few hours later, and finally some good news came through. Larso admitted he'd had reservations about continuing given the weather, his limited climbing experience, and the fact that everyone else had withdrawn. On the other hand, he continued, he was here to do this thing, wasn't the type to give up, and trusted us—Baz's ability and route planning more than anything—and, on that basis, he was up for it. Baz and I were thrilled: to do it as a three-man team was not only safer but was also the same number as Shackleton had in his team; moreover, Larso was good company and a tough customer. By coincidence, we even reprised the roles of Shackleton, Worsley, and Crean during the crossing as expedition leader, navigator, and mountain leader. Once again, an unexpected synchronicity seemed to be at play.

Baz talked Larso through the details of how we would work as a three-man team, with Larso seeming more comfortable with what we had in mind with each passing moment. I added the fact that Baz and I as nonsailors had entrusted Larso and Nick with our lives crossing the ocean and that if he could place a similar trust in us in return, then we would be set. He was more than happy to do this and the decision was made. We now needed Larso to ready himself for departure so that he could be up with us on the mountain within the hour when we decided to go.

The weather, however, was still not cooperating and we remained imprisoned in our tent by screaming sixty-knot winds, waiting for a break in which we could reasonably get going. The radio roused us again. Baz and I looked at each other, wondering what was coming next. It was the doctor calling to say that if Nick's feet deteriorated any further he might need evacuation to the Falklands. It was yet another body blow—so soon after hearing the good news from Larso, I once again realized the extent to which medical judgments could stop the whole expedition

Paul striding up from Possession Bay toward us on Shackleton Gap. He promised to keep his race face on.

dead in its tracks. That was the way it had been all the way through this expedition, though—two steps forward, one step back if you were lucky. From my standpoint, obviously if Nick really needed evacuation on medical grounds there would be no objection from me. What I doubted, however, was whether he did actually need it or indeed if he would be any better off in Port Stanley than he was under the medical supervision of our doctor, receiving anti-inflammatories in the warmth and comfort of *Australis*. I was, however, too tired to argue and, conscious that this conversation was most likely being recorded for posterity, accepted the news without emotion in my voice, hoping above all else it would not come to that.

Baz's talk of the military numbers game came back to me: the greater the number of participants in the expedition, the greater the chances of problems arising, any one of which could destroy everything. I was painfully aware too that we remained only a mile from the temptation of Possession Bay as an evacuation point and that this was eroding our chances of success. We knew we must get away from Shackleton Gap at all costs. Baz surveyed the scene outside. It was still a maelstrom but the wind gusts had diminished in their severity and the sleet had stopped. "Okay, that's it," said Baz. "Time to go." He wasn't wrong.

FALL LINE

11

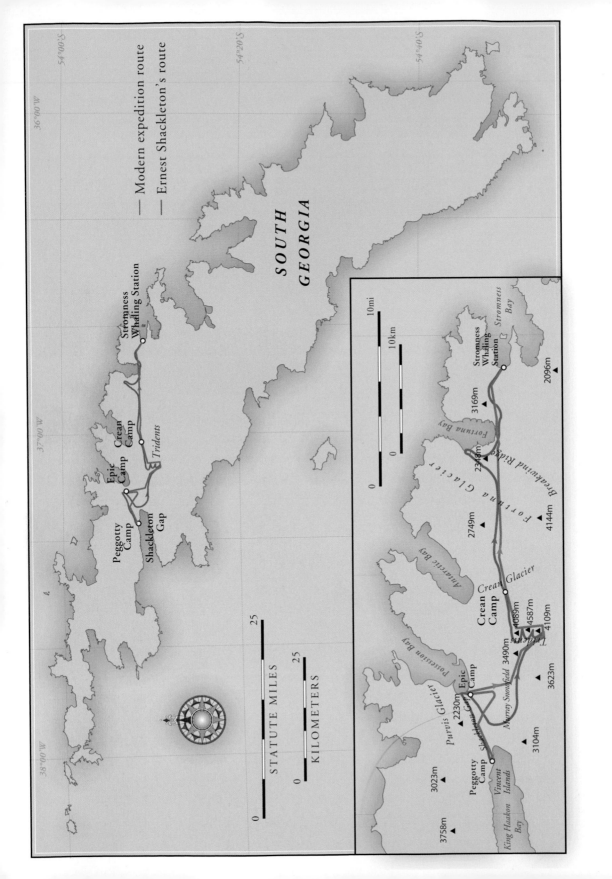

SOUTH
GEORGIA

Modern expedition route
Ernest Shackleton's route

Stromness
Whaling Station

Crean
Camp

Tridents

Epic
Camp

Peggotty
Camp

Shackleton
Gap

STATUTE MILES

KILOMETERS

0   25

0   25

Stromness
Whaling Station

Stromness
Bay

3169m

2096m

Fortuna Bay

2348m

Breakwind Ridge

Fortuna Glacier

2749m

4144m

Antarctic Bay

Crean Glacier

Crean
Camp

4089m

4587m

4109m

Tridents

Possession Bay

Murray Snowfield 3490m

3623m

Purvis Glacier

2230m

Epic
Camp

Shackleton Gap

3104m

Peggotty
Camp

Vincent
Islands

3023m

King Haakon
Bay

3758m

0   10km

10mi

"Fortitudine vincimus—By endurance we conquer."

Shackleton family motto

Three dark shapes appeared out of the mist 250 meters downwind of the tent. I turned to Baz: "They're here." "Okay, good," said Baz. "Let's get Larso in for a chat and then get cracking. I don't see this getting better so there's no point waiting around any longer." I agreed. This location had been bad in so many ways. Not only did it represent three kilometers of retracing our steps to end up in a very sticky spot, it had left the six-man team that departed Peggotty with such promise just two days earlier fractured by injury and self-doubt.

Now, however, we were down to a core of three men prepared to live with the risk of what lay ahead, knowing that support would not be able to reach us. I remembered Worsley's description of the conditions that often prevail in the mountains of South Georgia: "The hell that reigns up there in heavy storms, the glee of the west gale fiends, the thunderous hate of the grim nor-wester, the pitiless evil snarl of the easterly gales, and the shrieks and howls of the southerly blizzards with ever-oncoming battalions of quick-firing hail squalls, followed by snow squalls, blind a man or take away his senses." Not to be taken lightly, then. Just as well we had good clothes, I quipped to Baz.

Much of the time, life doesn't allow you to see the significance of decisions you make during those defining moments that we all experience. It is only later that you come to fully appreciate their consequences. In our case, removing ourselves from the danger we faced or taking the easy way out was, as ever, seductive, but we would not succumb to it.

Baz and I and now Larso could see the significance and danger of what we were about to attempt and with complete clarity decided we must go. We made

South Georgia and the trek ahead.

Previous pages: The spectacular but tortured landscape of the Crean Glacier in the post-dawn light. Baz leads our party of three.

the decision almost as if twenty-two men's lives on Elephant Island depended on it. If we had decided to quit, the bitterness of the decision would have remained long after the initial sweetness of removing ourselves from danger had gone. Luckily we all had the presence of mind to know that. We would carry on—for ourselves, for the others who remained on *Australis*, and to honor Shackleton's memory and the trust of everyone who had backed us to do this thing. It was time to go. Stromness awaited.

The tent door opened to reveal a bearded face in modern gear followed by a flurry of snowdrift. "Good to see you, Larso," we said, more or less in unison. "Good to be here," Larso replied before asking how we were getting on. "All good and ready to go," said Baz as we got down to the detail of how we would proceed. Baz would lead; I, as the heaviest man, would be the anchor at the back; and Larso would travel in the middle. All gear except the food we could carry in our small packs would go back down to *Australis* with Jamie and Joe, who had come to film our departure. Larso would remain in his modern gear and carry a tent and bivvy bag for emergencies. Baz and I, of course, would stick with our old gear, still damp from our soaking thirty-six hours earlier. We would all be roped together using a modern climbing rope, since that was what Larso was used to.

The delay for Si's evacuation and the subsequent loss of our backup team meant we now would be at the most dangerous section of the crossing, in among the

The last they would see of us until the end: Baz, Paul, and I disappear into the mist.

crevasses of the Crean and Fortuna glaciers, in complete darkness. On that basis, we agreed we would stop before we got there and put up the tent until dawn. It would be suicidal to travel in darkness with no lights or moonlight, as unlike Shackleton we were not blessed with good weather or clear night skies with a full moon to travel by. In Shackleton's words, "The friendly moon seemed to pilot our weary feet. We could have had no better guide." It allowed them to continue virtually without stopping, save for small breaks for reviving hoosh, hot milk, and short sleeps. Shackleton famously didn't sleep, however, for fear that they would never wake up again if they all slept, telling the others, whom he woke after just ten minutes, they had slept for a full half hour to boost their energy and morale.

Before we set off, interviews were filmed in the driving wind until cold forced a halt and reminded us that we needed to get moving. Jamie again asked us if we were okay to film things from this point on. As if there was much alternative! As we were breaking camp, my feet were getting hopelessly wet as five sets of boots had turned the area into a slushy mess. The toes on my right foot in particular were now numb, something I hoped would improve with our workload climbing up the pass toward the high ground of the Murray Snowfield. We moved off as the cameras followed us for a short distance. It was a relief to be on our way, the steepness of our ascent from our forlorn little camp, the strength of the wind, and poor visibility ensuring we were on our own within 100 meters. The three of us were in an uncompromising frame of mind. The next time we saw the film crew would hopefully be in Fortuna Bay, meaning the dangers of the mountains and crevasses would be behind us.

We ascended with temperatures hovering around zero but made far colder by the wind speed. The weather was pretty awful, but it was preferable to the intrigue and uncertainty of staying put, with circumstances beyond our control threatening constantly to destroy our dream and crush our efforts.

For the second time, we trudged through the deep snow of the Murray Snowfield in poor visibility. Last time it was in darkness; now it was through mist. A thought struck me in that swirling mist: our plan had been for eight men to cross the mountains, but the Southern Ocean and South Georgia's weather had seen off five. Now we had exactly the same size team as Shackleton and we even filled the same roles. He had done the crossing with Crean the strong man and Worsley the navigator. McNeish and Vincent were unable to continue and remained at Peggotty Camp, and McCarthy was required to stay behind to watch over them. As for our team of three, I regarded Baz as the Crean of our expedition, hard as nails and uncomplaining, while Larso had taken the sextant readings in rolling seas as navigator, just as Worsley had done ninety-seven years before. The parallels didn't stop there. Just as Crean ended up being the "Primus expert and chef" of the *Caird* crew, Baz had taken on cooking duties among the six of us and, like Crean, also regularly provided musical offerings. While Baz's songs were recognizable, Shackleton wrote of Crean's singing at the helm: "Nobody ever discovered what the song was. It was devoid of tune and as monotonous as the chanting of a Buddhist monk at his prayers; yet somehow it was cheerful." Which was preferable, I can't be sure. Inevitable comparisons would be made between Shackleton and me as expedition leader, in terms of our role at least. Our Jarvis-Gray-Larsen combination was the precise equivalent of the Shackleton-Crean-Worsely team, and it felt right at many levels that we should be the ones doing the crossing and with no support.

The going was tough, particularly after two days tent-bound with all of the crises that had unfolded. Having climbed up to a ridgeline, we now headed due east, knowing the bearing would mean at some stage we would hit the Trident mountain range. After four hours it still stubbornly refused to reveal itself until over the howl of the wind we heard the rumble of a distant avalanche ahead of us. It was nowhere near us, but it indicated that steeper ground lay ahead in the mist.

As we made progress toward the Tridents, the weather began to abate such that at 10 P.M., some five hours into our journey, the clouds cleared for the first time, giving us all one of the most spectacular views we'd ever seen. There was the jagged barrier of dark, pyramidal peaks that were the Tridents, emerging as they did out of the deep, white blanket of the snowfield. We were standing virtually directly beneath them, with our crossing point being about 400 meters to our east. Conditions underfoot were heavy going and, just to ensure against complacency, each of us broke through thin snow bridges covering crevasses several times, reminding us to keep the rope taut to prevent us from plunging any deeper.

The swirling mist came and went as we closed in on our target. Baz took a bearing to the right of the largest peak and the huge bergschrund, or main crevasse, that ringed it, giving both a wide berth in the knowledge that the saddle Shackleton descended so dramatically was close at hand. The Tridents have four possible crossing points in the form of saddles running north to south. According to both Shackleton's and Worsley's accounts, they tried them all from south to north, or right to left as you face them. The most southerly, lowest-looking saddle to the right as you approach from the Murray Snowfield was the most promising viewed

The Tridents, gatekeepers to the glaciers beyond. We descended down the pass between the third and fourth peaks from the left.

from the Murray but, in Shackleton's words, "the outlook was disappointing. I looked down a sheer precipice to a chaos of crumpled ice 1,300 feet below." The second saddle presented the same problem and so too the third and finally the fourth where they "could not see the bottom clearly owing to the mist and bad light, and the possibility of the slope ending in a sheer fall occurred to us; but the fog that was creeping up behind allowed no time for hesitation." The encroaching darkness forced their hand and they had no choice but to descend.

Today, the best descent route—unless you have ropes with which to rappel—is the third saddle. The fourth, which Shackleton went down, now had several exposed bands of rock at intervals on the way down that would be difficult to negotiate and impossible to slide over as he had, the protective covering of snow now gone. On our arrival at the third saddle, the view to the tortured rivers of ice of the Crean and Fortuna glaciers far below eclipsed even our first spectacular view of the Tridents only an hour before. With the improved visibility lower down we could see all the way east to the jagged peaks of Breakwind Ridge, still twenty kilometers distant.

The top of the saddle was a steep, wind-blown cornice of snow or, as Shackleton described it, a razorback. Immediately below it was a sixty-degree snow slope that was steep enough to mean that a fall, if not arrested, would send you tumbling toward a crevasse visible about 150 meters down and potentially pulling the others down with you. The crevasse's downslope face of turquoise ice had been forced upward, appearing like a low wall from our vantage point, meaning there was no way to avoid it if we fell. We avoided it by traversing the slope in a northeasterly direction, keeping one eye firmly on the placement of our feet and the other on the crevasse that lay in wait below. Each man leaned back into the slope, with my heavy carpenter's adze proving difficult to use given its weight and the ease with which the sharpened wooden handle buried itself up to the blade in snow each time I placed it. Increased effort was required to remove the three-kilogram ax each time, my arm burning with lactic acid.

The snow conditions on the descent were at least stable, allowing us a good footing—just as well considering the quality of our footwear, the leather of mine having long since become slippery mush, with the centimeter-long hobnails providing little purchase. Baz meanwhile wore boots with four-centimeter screws protruding down through the welt, which, although longer than mine, had begun to splay sideways, rendering many of them useless. Larso wore modern, plastic mountaineering boots and crampons, giving him good grip, although he, too, was well aware of how exposed we were. As a nonclimber, he coped fantastically well.

Once clear of the huge crevasse, we switched back due east, joining the fourth pass that Shackleton had descended about halfway down. Descending another hundred meters, we came across what appeared to be a relatively unobstructed, albeit steep, slope of perhaps forty degrees that went all the way down to the distant glacier. I commented that "ninety-seven years ago Shackleton would have come sliding past us at this point" just as Baz suggested that we, too, should slide to save time with darkness not far off. In true Shackleton style, roped together across the slope, we descended straight down the fall line on our backsides, finally coming to rest as the slope petered out and friction slowed us. In poor light this would have been extremely dangerous, as the odd large rock protruded through the snow. But in the reasonable visibility we had, we actually enjoyed the free ride and it was a thrill to be doing what Shackleton had done. By all accounts it was something of a thrill for Shackleton and his men too. "We seemed to shoot into space," Worsley recalled. "For a moment my hair fairly stood on end. Then quite suddenly I felt a glow, and knew that I was grinning! I was actually enjoying it. . . . I yelled with excitement and found that Shackleton and Crean were yelling too." Although he had no mountaineering background, Shackleton might have had an inkling that a controlled slide is sometimes safer than climbing down on foot, where one stumble can result in an uncontrolled tumble.

Within a minute we had descended hundreds of meters and were down on the edge of a huge snow bowl at the bottom of the Tridents looking back up at the steep slope. We were relieved to have gotten a major obstacle out of the way and impressed that we'd managed to descend in the fashion we had. As Shackleton said at the time, "We looked back up and saw the grey fingers of the fog appearing on the ridge, as though reaching after the intruders into untrodden wilds. But we had escaped."

The toes on my right foot were still resolutely numb as we made good time across the seemingly crevasse-free bowl, the ground to our north sweeping down to Antarctic Bay and ending in spectacular jagged ice cliffs. We were now completely committed, having descended the Tridents, and with the Crean and Fortuna glaciers and a climb up over Breakwind Ridge and then down into Fortuna Bay still ahead as darkness fell. Due to the delayed timing of our crossing, we made camp for a few hours until first light as planned, finding a

A steep climb down, followed by a glissade, or bumslide, awaits.

small circle of rocks seemingly designed for the task that we referred to as Crean Camp.

    I quickly removed my right boot to assess my foot, revealing toes that were startlingly alabaster white and completely dead to the touch. I massaged them for an hour and then donned warm socks, resting my mug of hot milk against my foot and subsequently wrapping the foot in the emergency sleeping bag. Three hours later, I still had no sensation. I was shocked at the insidious way it had crept up on me; quite honestly, I was shocked it had happened at all given my experience of cold weather travel. But this wasn't the dry extreme cold of the high polar plateau I was used to; these were wet, cold, constricted conditions wearing shoes that afforded no insulation.

    We broke camp, roped up, and surveyed the tortured landscape ahead, aiming to keep to the southern side of the glaciers in a line alongside the large mountain spurs that run into them. The ground underfoot was icy, making for better conditions in which to travel than the heavy snow of the Murray. Thankfully, the wind had also abated. The landscape was breathtaking, bathed in the orange glow of the early morning sun. Shackleton experienced the beauty of the spot in a different light. Having spent an hour in complete darkness at a spot near our Crean Camp, he noted: "a glow which we had seen behind the jagged peaks resolved itself into the full moon, which rose ahead of us and made a silver pathway for our feet.

Crevasses, deep, frequent, and deadly to negotiate with very limited gear.

Along that pathway in the wake of the moon we advanced in safety, with the shadows cast by the edges of the crevasses showing back on either side of us."

We traveled through a surreal, honeycombed labyrinth of fissures and holes, teetering on the icy ridges and crust above. Like an abandoned archaeological dig, trenches and subterranean features lurked below us as we picked our way through in silence, often on the lip of a mighty chasm that descended into darkness. At least the crevasses were visible. Had we tried this the night before we would not have managed a hundred meters. There was a complete absence of snow for the most part, the wind having whipped it away, and our route was exhilarating as we looked down into these deep fissures. Most worrying were the ominous snow sections every few hundred meters that, at perhaps fifty meters across, were reminiscent for me of recently formed Arctic sea ice. You hope it will hold your weight but expect it to crack at any moment and swallow you up. With ten- to fifteen-meter gaps between each of us on the rope, it could easily have done so.

We stopped to rest and surveyed the vast bowl of snow that lay ahead. We decided to light the stove, as our energy levels were depleted with the cold, concentration, and steady ascent of the past hour. The bowl was part of the river of glacial ice that spills down from the mountainous interior of South Georgia into Antarctic Bay to the north. This section between the Crean and Fortuna glaciers was unnamed but full of crevasse danger just the same, and we'd had multiple small falls with legs punching through into unknown voids below, and stopping for warm food boosted energy and concentration levels.

Our crossing of the Crean had been very different from Shackleton's. Temperatures for him were lower than we were experiencing due to both the time of year and climate change, which has resulted in the retreat of 97 percent of South Georgia's glaciers since his day. This meant that the Crean he would have crossed would have been more like the bowl that now lay ahead of us than the fractured river we had just traversed.

Our mood was good as we knew that this snow bowl, the Fortuna Glacier, and Breakwind Ridge were all that lay ahead of us if we could just keep our pace up. We needed to avoid another night out at all costs. Putting a tent up last night was forgivable since our timing and route had been dictated to us by the need to head off course to rendezvous with *Australis* to evacuate Si and the bad weather that ensued in that terrible location. To put the tent up again, however, would be straying too far from the spirit in which Shackleton accomplished the crossing.

I took over the lead of the next section, having developed a good feel for where crevasse danger might lurk from our experience thus far. The three of us threaded our way across the vast landscape toward the Cornwall Peaks, the steady pace allowing us to visit a place in our minds rarely visited as we journeyed along. For me, the vastness of such a landscape strips life back to basics and allows you to experience an almost meditative state—the core of who you are that the normal pace of life seems to mask from view.

Perhaps an extension of this is what Shackleton spoke of when he mentioned the presence of another with them on the crossing, a phenomenon confirmed by the two other men: "When I look back at those days I have no doubt that Providence guided us, not only across those snow-fields, but across the storm-white sea that separated Elephant Island from our landing place on South Georgia. I know that during that long and racking march of thirty-six hours over the unnamed mountains and glaciers of South Georgia it seemed to me often that there were four, not three." Unprompted, Worsley said to Shackleton a few days after the crossing, "Boss, I had a curious feeling on the march that there was another person with us" and Crean "confessed to the same idea." Was it the presence of another? Or was it a part of their being that they were almost unaccustomed to experiencing? I had asked myself the same question many times.

We looked out over the same scene Shackleton had all those years before, but now these "unnamed mountains" were called the Cornwall Peaks, named only a few years after Shackleton's death by the 1920s Discovery expeditions, whose goal was to look at the viability of whaling. They marked the next of Baz's predetermined "report lines"—locations where Larso would call *Australis* on the satellite phone to tell them where we were. Reaching the Cornwall Peaks meant we had successfully crossed the Crean and that the Fortuna and Breakwind lay ahead of us.

We ascended quite steeply to a saddle between two nunataks—a smaller one to the north and the main 840-meter-high Cornwall Peak to our south. The climb had been tiring and we stopped to survey the wide sweep of snow and ice of the Fortuna Glacier we now needed to cross. Unlike the Crean, all of its dangers remained hidden from view. We decided to head east, aiming in a straight line for the last nunatak on the far side of the Fortuna some three kilometers distant. We knew that around that corner lay an ascent to another saddle that would finally lead down Breakwind Ridge into Fortuna Bay.

Baz, now somewhat rested from his spell at the rear, resumed trailbreaking as we dropped into the wide bowl the mighty Fortuna occupied. The snow seemed

Top: Looking back over the hidden crevasses, the steep descent of the Tridents we'd just come down marked in the distance.

Bottom: The scenery was spectacular but dangers lurked just below the surface.

solid enough, leaving us to think that our passage across might be a safe one. Suddenly Larso, 300 meters into the crossing, dropped in to his waist, reminding us of what lay beneath. Thereafter each of us had small falls, usually just a leg breaking through, although Baz and I sank to our waists on a couple of occasions. With our senses on high alert, we longed to reach the other side and be free of the crevassed terrain we'd been negotiating since we broke camp almost ten hours earlier.

We approached the nunatak painfully slowly, as is always the way in the polar regions, where distances are vast, travel on foot slow, and the air so clean and clear that it makes distant objects seem closer than they really are. The final approach to it revealed a bergschrund around its base. We decided to give this a wide berth, turning north to follow the Fortuna down toward the coast in the hope that any crevasses that might await us there could be crossed at right angles. This would allow us to step over them until we got down to a point where we could again turn east and head up the next pass to the saddle and then descend Breakwind into Fortuna Bay. It was not to be. Baz plunged waist-deep into a crevasse, the lip of which was soft snow that threatened to collapse at any moment. He bridged the gap with his arms, kicking out for purchase with his feet before finally finding a wall he could push off to extricate himself as Larso lay on his ax and I on the adze providing an anchor. Quickly composing himself, Baz walked back to the west around the crack that indicated the line of the crevasse. When he was twenty meters beyond it, he felt safe to move north back down the Fortuna but disappeared a second time, this time more dramatically than the first. He was only visible from the shoulders up as I lay prostrate on the ground, the flat head of the adze buried deep in the snow. Baz reappeared but not without a few choice words. Among other things, he cursed the decision to coil the rope diagonally around our bodies rather than retain the improvised sling we'd made from the hemp rope that had proved too uncomfortable to walk in. The coil had virtually strangled him, but we were through, and got a brew of hot milk going while we assessed the next steps. The Boss would have approved.

The wind picked up at our backs, bringing with it rolling clouds of fog on the far side of the Fortuna and the distinct feeling that the mountains were not yet ready to release us. The mountains of Breakwind Ridge ahead formed a formidable barrier, the obvious chink in their armor being a more rounded, lower peak that lay more or less in line with the easterly bearing we'd taken across the Fortuna. With the cold nipping at our heels and anxious not to be enveloped in the advancing clouds, I led as we ascended a broad bowl several kilometers across

The edge of the world: a near-vertical descent awaits us over the edge of a serac, high on Breakwind Ridge.

to reach a saddle between the rounded peak and its more angular neighbor. We had reached the high point of our journey in every respect, and all that we now needed to do was concentrate on the descent into Fortuna Bay. We celebrated but did so prematurely: as we crept toward the edge of the saddle, it revealed itself to be a knife-edge, corniced ridge with a very steep snow slope beneath it leading to a second lip 100 meters below. Beyond that we couldn't see, and for all we knew this second lip could end in a sheer rock or ice cliff of unclimbable steepness.

Ironically, Shackleton had found himself in a similar situation after an initial detour. Having turned to the northeast too early in the moonlight and lured on by what Worsley had initially described as an easy descent, he headed down an arm of the Fortuna that ended in a precipitous, hanging glacier front high above the sea. "We must turn back for a while," announced the Boss, sounding as matter of fact about the disappointment as he could as they turned around and ascended the slope. A mile on they turned left to head south and follow the line of mountains that make up Breakwind Ridge, all the while looking for a way down. Worsley describes how they "were making for the only opening in a ridge of rocky mountains that lay athwart [their] course." Most modern-day adventurers have interpreted this as the gap at the southern end of Breakwind Ridge, as we had done ourselves until now. However, on approaching from the west, as we were, the logical gap in Breakwind seemed to be this lower, rounded mountain and the pass next to it that lay ahead of us to the east, luring us on.

Shackleton's Toothgap Ridge descent, according to Worsley, "was precipitous." He continued: "It went off so far as we could tell into a sheer cliff but to the right it looked as though there was a possible descent." With the help of the ship's chronometer, they stopped a way into their descent, knowing the time approached 7 A.M., the hour when work would commence at Stromness whaling station. Sure enough, a whistle sounded as 7 o'clock arrived, calling the men to work. "Never did music sound so sweet to our ears as that whistle," Shackleton commented. It was still a good distance away but manageable save for the dangerous descent they were committed to. For us the lure was the lush green of the valley floor 500 meters below, and then Stromness, but it was still a world away until we'd negotiated our formidable descent.

Shackleton's team went down a dangerous slope, its steepness not becoming apparent until it was too late. Said Worsley: "A single slip from one of us would have meant the end of all three." To get down, Shackleton cut steps with the adze initially and then "walked downhill," lying flat on his back and smashing

Baz surveying the scene ahead once he knows I have arrested my fall. I went from last in line straight past Larso, who took this shot seconds later.

steps into the ice with the heels of his boots, which the other men then followed. They descended about 1,000 feet to the shore for what Worsley described as "fifteen minutes of splendid tramping over a level beach." He then went on to describe coming "to the front of the great glacier which fortunately did not quite reach the sea, where we crossed long gravelly flats of the glacier almost like quicksands in which we sank half-way to our knees. The going was good for half a mile along the beach at the head of Fortuna Bay." From our vantage point we could see that different terrain awaited us, with the ravages of climate change having withered the König Glacier so that now its snout was many kilometers from the beach. Our immediate challenge, however, was to descend Breakwind.

The parallels were striking between the two climbs down off Breakwind. Like Shackleton we assessed our options, deciding to commit to our descent, too tired to retrace our steps or try to find another way down. Plus we thought this route resembled very closely the path taken by Shackleton, referring as he does to being lowered down slopes as steep as a "church steeple" and cutting steps with the adze. We at least had more snow rather than sheer ice, allowing us to turn and face the seventy-degree slope, kicking steps as we went and leaning our faces into the wall to stop ourselves falling backward. Only when the steepness abated momentarily did we finally turn to face the incredible vista of Fortuna Bay.

Kicking steps in the snow with our heels and, for me, planting the adze was desperately hard work, and, although the snow was quite deep, the soles of my boots were slippery. Larso fared much better with his crampons and ax and Baz, like me, somehow managed in his awful boots, a testament to his superior climbing skills. As we descended, we would pause every twenty or thirty meters to catch our breath and for me to release the lactic acid in my arms from carrying the heavy adze. Each time we stopped, Baz would try to assess the way down as the slope was too steep for us to see the bottom or indeed see if there even was a way down.

About halfway through our ordeal, we realized we had descended into a chute that ended in a cliff that was too steep to descend any farther, forcing us to ascend twenty meters back up the steep snow slope and traverse across a ten-meter rock face. The drop was sheer as each of us skittered across, although despite the exposure, I actually felt more comfortable on the rock as I could at least wedge the points of my hobnails into the nooks and crannies to get purchase, something I couldn't do on the slippery snow surface. We all knew that, as for Shackleton, a fall from here would likely have pulled us all down and over the edge of the cliff and meant the end of all three.

Baz looks for a way down off a rocky ledge—a brief respite in the steep and dangerous descent off Breakwind Ridge.

Luckily the traverse proved worthwhile, and below us now was another steep snow slope, followed by a low vertical rock band that looked climbable if we were careful. Guiding one another's feet and hands into good hand- and footholds, we got down the final rock band, arriving at a ledge before a scramble sideways onto more climbable ground. Two hundred or so meters later, we were down to the valley floor. I'd always assumed that Breakwind had been named by Shackleton based on its forming a natural barrier to the westerly winds that had been at our backs the whole way, but, given the extreme nature of our final descent, I'm not so sure. As far as I was concerned, it was more "Touch Cloth" than "Break Wind."

We were greeted by the noise of penguins and shortly thereafter by the film crew, full of questions about how it had been, and how excited we must now be. We were, of course, excited but, to be honest, were still reliving the recent descent down Breakwind and were focused on trying to cross the surprisingly substantial meltwater barrier of the König Glacier. The toes on my right foot were still numb and I certainly wasn't keen to soak my boots with a wade through the raging torrent of meltwater that surged through a break in the shingle beach from the glacial lake dammed up behind it. Seals were struggling to swim against the current and I didn't fancy trying my luck, especially with eight kilometers still to travel and with night gradually closing in. I also knew it would be foolhardy to remove my boots, given how numb my right foot was.

Hobbling in pain, I was preoccupied with wanting to see what state my feet were in—particularly the numb toes on my right foot. But I wasn't going to do it with cameras present. Baz, meanwhile, was tired of being stuck on the wrong

Past Crean Lake, which almost claimed Shackleton's party at the last hurdle when they plunged through its icy surface. In the distance the arrow shows our descent off Breakwind.

side of the meltwater channel and subjected to questions, and so retraced his steps to the edge of the lake. Casually gauging its depth based on a crude assessment of how far out of the water a distant penguin seemed to be protruding, he strode biblically forth with Worsley-esque abandon. The penguin had in fact been standing on the far shore but, luckily for us, the water was only knee deep.

Ferocious katabatic winds gusted down the valley, their approach easy to see, preceded as it was by disturbance on the water and penguins being bowled over like tenpins. With wetter feet than normal, I didn't pause for the camera crew, instead heading for the far side of the valley to assess my right foot. I climbed the steep slope following the band of darker vegetation that marked the flow of the small stream out of Crean Lake as the most direct route up the slope. Shackleton would have done exactly the same thing, explaining why he and Crean ended up walking across the frozen lake and plunging through the surface. Managing to stop to wring out my socks and massage my ivory toes before Joe the cameraman bounded into view for an interview, I got moving again as dusk and cold came on quickly.

Walking over the strange lunar landscape of uneven rock was desperately painful. Sparks from my boots were a thing of the past as many of the hobnails had either gone or been bent sideways or driven up into the soles of my boots with the impact of walking on the rocks for which they were not designed. When we finally reached the top of the pass, a roaring wind at our backs afforded us our first view of the prized Stromness, five kilometers away and rust red in the late evening sun. It was long since deserted; the only movements far below were

the glints of light on *Australis*'s superstructure as it scudded across Stromness Bay, headed for a rendezvous with us at the deserted whaling station.

We regrouped at the top of the pass, congratulating one another warmly for our combined focus and concentration of the past four days. We kept moving for fear of stiffening up and not being able to get going again, descending the steep, loose rocks while doing our best to preserve our feet as they slid forward in our boots.

On the final approach to Stromness I soothed the burning, raw sensation on the base of my feet by walking through the stream on the valley floor, thankful that I didn't need to preserve my boots any further. Larso and I sat to wait for Baz, who had stopped to look at the waterfall down which Shackleton and his men inexplicably descended, perhaps due to extreme tiredness and the fact that the ground around them was covered in snow. Otherwise I saw no reason for them to either need or want to descend it.

Larso closed his eyes and had a microsleep. When Baz appeared five minutes later, I woke Larso, telling him he'd had half an hour in a parody of Shackleton's motivational strategy of allowing Worsley and Crean to think they'd slept for longer than they had. He laughed knowingly, scarcely able to believe like me that around the final knoll lay the promised land of Stromness, once we'd run the gauntlet of aggressive fur seals—just a short stagger away.

Ghostly white figures—not the ghosts of whalers but our colleagues, wearing the conspicuous white of our modern Henri Lloyd overalls—stood waiting to greet us at the water's edge. We had taken ninety-six hours to Shackleton's thirty-six as night fell on the longest of days and on an expedition that at times I'd thought would never end.

NEVER THE LOWERED BANNER

12

Sound, sound the clarion, fill the fife!
Throughout the sensual world proclaim,
One crowded hour of glorious life
Is worth an age without a name.

Thomas Mordaunt, *The Call*

W*alking into Stromness in the failing light of 11 January meant that we achieved the "Shackleton double," bringing with it a mix of emotions that are hard to describe. Certainly there is elation, relief and pride and a great sense of camaraderie amongst the team, but added to this there is now an overwhelming sense of feeling humbled by our achievement. Without doubt this relates to having got closer to understanding what Shackleton went through on the original journey almost a century ago. The pain, fear, suffering and doubt that he and his five colleagues needed to overcome to achieve their incredible journey of survival as winter approached, [while] the majority of his men remained behind on Elephant Island clinging to life and no backup existed, simply because no-one knew that he or his men were still alive. All of this after enduring a year and a half on the crippled* Endurance *and the floating pack ice of the Weddell Sea. That we have managed to emulate some of this story and get close to the kind of determination he needed to win through is truly humbling.*

Old gear, new challenges: me, dressed for action.

Previous pages: Preparing to re-create history.

I wrote these words immediately after finishing our journey and I stand by them now. Without doubt they go a good way toward describing the way I feel. With hindsight and my right foot now recovering, however, I can look afresh at the events of our two expeditions nearly a hundred years apart and draw some conclusions as to how Shackleton triumphed through such adversity and why he was a restless spirit constantly drawn to the next challenge. Speaking to Worsley on board the *Caird*, Shackleton put his life philosophy simply: "Never for me the lowered banner, never the last endeavour."

There were certainly many interesting parallels and some clear differences between our two boat journeys, although obviously our intention was to replicate what he'd done as precisely as possible. Setting out obvious differences between them enables a clearer assessment of his methods and motivations from the body of our common experience.

By way of similarities, the *Caird* and the *Alexandra Shackleton* negotiated a band of brash ice that orbits Elephant Island (according to Worsley, pronounced "Hell-of-an-Island" by the men), traveled past icebergs, and were accompanied by whales and albatross. All the while we both remained mercifully untroubled by pack ice during our journeys—something we were particularly relieved about given that in 2013 Antarctica's pack ice was at its maximum extent for decades.

In terms of differences, Shackleton had set off with a team whose goal was to cross the Antarctic continent in order to eclipse the achievements of all who had gone before—a journey that perhaps would be Shackleton's swan song. As such, his team of twenty-seven included surgeons, a cook, an artist, engineers, scientists of various disciplines, proven land-based expeditioners, and, of course, seamen of the highest caliber, all recruited for their skill and resilience along with other, more novel traits. Our six-man expedition team, meanwhile, reflected skills required for just the small boat journey and climb over South Georgia—a journey that we all specifically signed up for but that none of the *James Caird* crew could have ever anticipated doing.

That our two teams departed from Point Wild after very different journeys to get there is taken as a given. Ours was a journey of organizational challenges but relative safety; Shackleton and his twenty-seven men had endured a brutal, dangerous, cold, five-day open-boat journey in their three lifeboats across the sea to Elephant Island from the edge of the pack ice. This followed ten months' incarceration aboard their stricken ship and five months camped on the ice that is an epic tale of survival in its own right and one we could not, and did not, experience. It meant that we set off from Elephant Island in a very different frame of mind. Regardless of the difficulty of the journey ahead for the *James Caird*, I feel Shackleton's team was on an improving trajectory psychologically after the terrible disappointment of the failure of their original goal to cross the continent. After all, they had at least escaped their dying ship, survived the pack ice and the open-boat journey to Elephant Island, and could build on what had already been achieved. We, on the other hand, had come from a position of safety and were thrust into our challenge, no pun intended, cold. Common to both crews was the dangerous journey ahead, which, regardless of the fact they left through necessity and we voluntarily, aligned very closely once we were under way.

Our departure date, however, was January 23 whereas for Shackleton it was April 24—summer versus autumn and less cold for us as a result. We had no option on this; it was the only time of year we could get permits based on there being at least some other ship traffic in the event things went terribly wrong for us or *Australis*. Practically speaking, this had a couple of implications. First, it meant that Shackleton had air temperatures probably 5°C colder than we did (-5°C to -10°C for him and zero to -5°C for us), although the absence of a thermometer on his trip means we can't be sure. Sea temperatures would have been about the same, but the colder air would have made their lives more miserable, and also resulted in thick ice forming on the deck and sails of the *James Caird*. We only had occasional snow and sleet adhering to our sails and coating our deck in a thin layer of slush. Put simply, we didn't have the same problems of icing up, which made the *Caird* far more unstable than the *Alexandra Shackleton* through

Left: Baz, who led the land crossing of South Georgia.

Right: Shackleton's Polar Medal— recognition of his incredible achievements.

a combination of extra weight on deck and reduced weight below due to their being a lighter crew by perhaps 100 kilograms—the weight of a large man. The fifteen inches of ice on the *Caird*'s deck that both Shackleton and Worsley mention in their accounts could, even if they overstated it, have easily equated ten or perhaps twenty times the weight of our life raft and cameras on deck. This gave the *Caird* a "tendency to capsize," as Worsley delicately put it. The ice also meant the *Caird* sat lower in the water, making her more susceptible to waves inundating them. Some consolation was the fact that the men were physically smaller than us so had more room below deck. The additional presence below deck in the *Alexandra Shackleton* of safety gear including survival suits, life jackets, flares, and electronics meant we were even more cramped than they were, albeit theoretically safer.

In terms of the construction of the *Alexandra Shackleton*, we were as faithful as possible to the *Caird* with the exception that we put in an aft bulkhead, to separate the cabin from the open cockpit, and accessed the main area below deck via a hatch. This was part of our safety plan in that the air in our living area

One boat, six men, one hundred years: the James Caird *launching (left) and our re-creation of the moment.*

would be our main buoyancy aid in the event of capsize. The other difference was that while the *Caird*'s bow was completely watertight as far back as the main mast, from there to the stern she was decked with a lattice frame structure of four sledge runners and packing-case lids nailed together with canvas stretched over them. We used tongue-and-groove pine planks covered with cotton canvas that certainly leaked when waves crashed over us, but it was nevertheless better than they had.

We also had a support vessel, *Australis*, under the expert captaincy of Ben Wallis, and with a great crew in Magnus and Skye, which provided a safety net, although how much of one is difficult to gauge. Had a man fallen overboard in a storm, poor visibility, at night, or, worse still, in some combination of these, it would have been the end for him. Besides, the kind of weather that could cause such an event would likely have seen *Australis* preoccupied with her own survival and in no position to help us. Her very presence, however, was a temptation that Shackleton didn't have—the double-edged sword faced by modern-day explorers of knowing that the possibility of rescue exists. For Shackleton the

The James Caird *setting off from Elephant Island in 1916, with the* Stancomb Wills *at left.*

fact that he was alone on the sea with twenty-two men left behind on Elephant Island gave him a unique motivation to succeed.

Add to this the fact that the agendas of film and media, often conflicting with our own, impacted team solidarity, and that we needed to force-fit our weather with trying to complete his journey, and there were problems we had that he did not. This issue was brought into sharp relief when we tried to land at King Haakon Bay, a bay we would not have chosen to land at given the weather we had the day we arrived at South Georgia but did so in order to follow his route.

Shackleton and his men spoke of enduring gale-force winds on all but five of their days at sea, and of a "hurricane" on day sixteen. If these definitions are taken literally, a "gale" means seas of 5.5 meters high and winds of thirty-four to forty knots. A "hurricane" translates to sixty-four to seventy-one knots and seas of fourteen meters. We alternatively had gale-force winds during our two-day storm and some violent cross-seas with strong breezes and near gale-force winds on several other days but no hurricane. It is testament to their skill that they managed to survive such conditions with constant diligence required to ensure the boat did not broach on the waves and become swamped or capsize.

Our land crossing of South Georgia was fraught with issues. Testament to how damp and cramped conditions were below deck on our boat journey is the fact that half our team got trench foot, as did they. Worsley described it as "superficial frostbite" from the "constant soaking in sea-water with the temperature at times nearly down to zero, and the lack of exercise." Our six days waiting at King Haakon Bay for suitable weather, which resulted in three of our crew being forced to abandon the climb, matched closely the week Shackleton spent at King Haakon Bay—albeit in two camps, Cave Cove and Peggotty. And in incredible synchronicity, the three of us who remained fit enough to make the land crossing were the navigator, skipper, and expedition leader—just as it had been with his team.

They left at 2 A.M. and took just under thirty-six hours to complete the crossing to Stromness. This despite wrong turns down to Possession Bay, a detour up the Briggs Glacier, multiple attempts to get through various passes down through the Tridents, and a detour to Best Peak on the Fortuna Glacier that forced a backtrack along Breakwind Ridge until a suitable way down was found. We, on the other hand, went due east with minimal detours save for the need to evacuate Si the cameraman down to Possession Bay—a route that interestingly saw us follow almost exactly in Shackleton's footsteps on the detour he took. For our part, the crossing of South Georgia cost us five of our complement of

eight men in conditions that were frankly among the worst South Georgia has to offer, as opposed to the good weather he received, in a journey that took ninety-six hours to complete.

Shackleton set an incredible pace across South Georgia, taking uncharacteristic risks in order to get across and down whatever lay in his path as quickly as possible. However, he crossed in early May, when temperatures were colder than they were for us, by perhaps five degrees given the season and the warming effects of climate change. They also had very good weather with no rain or snowfall and a full moon to travel by while the ground over which they traveled was also less crevassed—a result of both the season and the fact that climate change has ravaged almost all of South Georgia's glaciers in the intervening ninety-seven years. In short, there were fewer crevasses and those that there were would have been covered with more snow when he crossed. To suggest that Shackleton, Worsley, and Crean managed this journey with no climbing or trekking experience also does them a great disservice. Shackleton had been knighted for getting closer to the South Pole on foot than anyone before him and Crean was perhaps the physically strongest polar explorer of his generation, both of them having had a great deal of exposure to crevassed terrain. Nevertheless, the speed of their crossing and the fact that they made it at all are both truly amazing given the paucity of their gear and the fact that the interior of South Georgia was completely unexplored.

What then can be said about how Shackleton achieved this journey against such incredible odds? Today's world may be more bureaucratic and media-driven, but I have no doubt that Shackleton would have been able to organize an expedition in the modern era. His entrepreneurialism and media savvy would have shone just the same. As Hurley observed: "Any day you may see Sir Ernest, always alone, taxi-ing from one newspaper office to another. He is trying to arrange the best terms and it is going to be a battle royal both for the news and pictorial rights." In a manner more modern than Edwardian, he knew the commercial worth of his ventures.

How our different journeys to Elephant Island affected our two teams is impossible to calculate. Shackleton and his men would have been worn down by but also well acclimatized to their surroundings, and certainly Shackleton's resolve would have been steeled to salvage something from the wreckage of

The König Glacier. Fully formed in Shackleton's day, it's now a victim of climate change.

both ship and expedition. The absence of any possibility of rescue also provided a unique motivator for the *Caird* crew. If they hadn't made it, the men they left behind would have perished, failure for them quite literally not being an option. It was this outlook Baz and I sought to emulate when tent-bound after Si's evacuation on the Shackleton Gap. As Amundsen said in his tribute to Shackleton in the *Daily Chronicle*, "The word failure is not found in Shackleton's vocabulary, and he succeeded."

That success is testament to Shackleton's ability to take constant problems in his stride; the ease with which he adapted to changed circumstances; and his determination, selflessness, and irrepressible optimism—in short, his leadership. This is reflected in the team he chose to take with him in the *Caird* and his determination to pursue the goal of saving all of his men with the same dedication, endurance, and good humor he had brought to the original goal of crossing Antarctica. To achieve this he led from the front, never asking another to do what he would not do himself and, more often than not, doing others' work for them. According to Worsley, Shackleton would always undertake the "most dangerous and difficult task himself. He was, in fact, unable by nature to do otherwise. Being a born leader, he had to lead in the position of most danger, difficulty and responsibility." His watches on the *Caird* as they journeyed to Elephant Island from the pack, with him going without sleep for three days, and his again resisting sleep during the crossing of South Georgia so that Worsley and Crean might rest, are legendary.

Shackleton's caring for others speaks much of the instinct for compassion which he had inherited from the women in his life (his mother and eight sisters in particular) and strengthened by a desire to treat others better than he had been treated as a young seaman in the merchant navy. Again Worsley seems to have summed it up best:

> Looking back on this great boat journey, it seems certain that some of our men would have succumbed to the terrible protracted strain but for Shackleton. So great was his care of his people that, to rough men, it seemed at times to have a touch of woman about it, even to the verge of fussiness. If a man shivered more than usual, he would plunge his hand into the heart of the spare clothes bag for the least sodden pair of socks for him. He seemed to keep a mental finger on each man's pulse.

He was tough but compassionate, finding ways to connect with each man even if he did not naturally identify with him. His decision to play cards with Hurley on the pack ice each day was a prime example. It gave the two men a reason to talk, where otherwise words between them did not come easily. He was, as an early shipmate said, "a Viking with a mother's heart"—a characteristic that made those with him extend themselves in the knowledge that he had their best interests at heart and that dying for king and country was not the ultimate goal.

His team was, of course, of the highest caliber, and under Shackleton's leadership they worked as one against the elements. The leadership and optimism of Shackleton and the skill of Worsley, McNeish, and McCarthy, combined with the strength and resilience of Vincent and Crean, meant they were the right men for the job. To my mind McNeish was unjustly maligned, his early insubordination on the pack having blotted his copybook with Shackleton. From this he would never recover in Shackleton's eyes, to the extent that he was controversially denied the Polar Medal along with Vincent, who had supported his challenge to Shackleton's authority. Still, his role should not be underestimated. Shackleton had got him to begin preparing the *Caird* for a long sea journey as far back as Ocean Camp, so plenty of thought and labor went into making her right for the job. That she survived was attributable in equal part to McNeish's skill and Shackleton's forethought. Even so, McNeish's efforts only gave them seventy centimeters of freeboard and had them bailing for their lives when waves crashed in and chipping ice from a deck so glassy that it is amazing no one was lost overboard.

And what of Worsley's near-mythical feat of navigation to reach South

True moral courage: Shackleton and Wild at Ocean Camp.

Georgia with only two or three accurate sun sights? Our journey has proved it can be done but only together with razor-sharp accuracy in dead reckoning for the remainder of the time—a critical point. You need to judge speed and direction accurately in between such sightings to ensure you remain on track and that cumulative error does not result in you either running into or missing South Georgia by the end. No one can diminish the enormity of Worsley's achievement, but huge credit here must also be given to the skill and dedication of our principal sailors, Larso and Nick. In Worsley's words, by day thirteen, "Since leaving Elephant Island I had only been able to get the sun four times, two of these being mere snaps or guesses through slight rifts in the clouds." By curious coincidence Larso also managed just two good noon sextant readings—on day four and day nine—with two other, unreliable, observations on days five and eight. These readings were less reliable due to their not being noon sights and given the conditions in which they were taken.

I am proud to have been involved in a journey where a combination of Nick and Larso's world-class sailing skills and the determination and focus of the rest of the crew got us to our target against such odds. It was an incredible feat and worthy of great praise. Having said that, they would agree that Worsley achieved what he did in rougher sea, and with pressure on him that we simply did not have. For me, the amazing thing about what he did is not the number of sightings he took of "Old Jamaica" to find South Georgia, but the circumstances in which he accurately took them. This can never be repeated.

And then there was their rogue wave. Near-mythical creatures until recently verified by measuring gauges on North Sea oil platforms and found to be real, rogue waves are defined as waves twice the height of anything else in a given sea state. We didn't experience one, although once in a while a wave that was noticeably larger than the rest would appear. What caused the one that struck the *Caird*, no one knows. It was certainly not a shallowing of the ocean: where they encountered it on day ten, 444 miles from South Georgia, the ocean is 3,000 meters deep. Could it have been a wave caused by a capsizing iceberg, as Worsley suggested? It is possible, but, given that they hadn't seen an iceberg for many days and were in such deep sea, we can only speculate. It is perhaps more likely to have been a combination of the giant rollers that pulse around Cape Horn from the Pacific—Worsley's "leviathans of the deep"—combined with seas generated by the southwesterly gale that had been blowing hard for three days from the same direction. Whatever the case, all we know for sure is that, as unlucky as they were to have experienced it, they were even luckier to have survived it.

Regardless of Shackleton's leadership, his team, and the skill and resilience they exhibited in worse weather than we experienced on the ocean, he did also enjoy a lot of good luck. I once heard it said that life is about "playing a bad hand of cards well." If so, Shackleton was the master, riding or making his luck in equal amount. A slightly bigger wave, a man falling overboard, lower gunwales, fewer views of the sun, the hurricane lasting a moment longer—all could so easily have resulted in epic failure. As Shackleton himself said, "I have marvelled often at the thin line that divides success from failure and the sudden turn that leads from apparently certain disaster to comparative safety." Judgment, skill, and determination certainly allowed them to prevail, but to say luck played no role in things would be wrong.

Much of what we might ascribe to luck today was for Shackleton a firm belief that "Providence"—or "Old Provvy"—wanted them to make it—almost a belief of Shackleton's that it was his destiny to do so, as much as any denominational religious belief. Interestingly, although he came from a Quaker background, he did not hold religious services on the *Endurance*. Among the mountains of South

Georgia, however, all three men firmly believed that another walked beside them, something from which they all obviously derived tremendous strength. Shackleton later wrote:

When I look back at those days I have no doubt that Providence guided us, not only across those snow-fields, but across the storm-white sea that separated Elephant Island from our landing place on South Georgia. I know that during that long and racking march of thirty-six hours over the unnamed mountains and glaciers of South Georgia it seemed to me often that we were four, not three.

It was not only the expeditioners themselves who were moved. T. S. Eliot, inspired by Shackleton's journey, wrote in his poem "The Waste Land":

Who is the third who walks always beside you?
When I count, there are only you and I together
But when I look ahead up the white road
There is always another one walking beside you.

This feeling continued after their ordeal was over. Within twenty-four hours of their arrival in Stromness, a gale swept over the island. According to Worsley, "Had we been crossing that night nothing could have saved us. The Norwegian whalers afterwards told us there was never another day during the rest of the winter that was fine enough for us to have lived through on top of the mountains. Providence had certainly looked after us."

Certainly then the strength of Shackleton's team, his leadership, the unique motivation that death as an alternative provides, and both luck and spiritual guidance all played a part in Shackleton's success. But perhaps the key reason he was able to motivate himself and others to pull off this journey against such incredible odds is because it represented precisely why he was in Antarctica in the first place: to pit himself against the greatest challenges he could find in order to discover more about what lay within him and us all. Shackleton said as much when writing after the expedition about arriving in Stromness with virtually nothing save their ship's log, adze, and cooker:

That was all, except our wet clothes, that we had brought out of the Antarctic, which we had entered a year and a half before with well-found ship, full equipment

and high hopes. That was all of tangible things, but in memories we were rich. We had pierced the veneer of outside things. We had suffered, starved and triumphed, grovelled down yet grasped at glory, grown bigger in the bigness of the whole. We had seen God in his splendours, heard the text that nature renders. We had reached the naked soul of man.

"We had seen God in his splendours . . . "

The need to rescue his men against seemingly impossible odds served this just as well as the original quest to cross Antarctica and meant he was able to bring all of his energy, optimism, and belief to bear on achieving it.

Although articulated considerably less eloquently than Shackleton's, my reasons are consistent with his in that I also undertake journeys to such places for reasons of self-discovery and to experience a greater understanding of life at some level. Antarctica, removed as it is from the noise of everyday life, allows you to explore who you are without having society dictate it to you. It is a challenging environment but in response to it a resourceful side of your personality emerges—one that only appears in response to challenging situations. Perhaps the third man "who walks beside you" is in fact the one who walks inside you—the one you are so unaccustomed to experiencing that it feels like another whenever you encounter it. Whoever or whatever it is, you feel drawn to come back and reacquaint yourself with it at every opportunity.

It was, then, far more than ego and opportunism that drove Shackleton. He regularly used the power of heroic imagery and poetry to motivate both himself and his men to achieve great things, not as a contrivance but because he believed in it and what it represented. The explanation he gave to biographer Harold Begbie for his love of the poetry of Robert Browning reveals a great deal about his own philosophy:

> But I tell you, what I find in Browning is a consistent, a spontaneous optimism. He never wobbles. You never catch him doubting a purpose in creation or quailing before the infinite. The bigger the universe, the more he likes it. He can't feel at home in the longitude and latitude of finity. There's no parlour scepticism in his soul. His spirit goes up with something more than confidence to meet the mountain crags and the stars. He loves greatness and vastness. It's the Whole that he is after, and the part can't trouble him. If he looks at doubt it is to smile, never to sigh. No poet ever met the riddle of the universe with a more radiant answer. He knows what the universe expects of man—courage, endurance, faith—faith in the goodness of existence. That's his answer to the riddle.

Like Browning—whose lines "I hold that a man should strive to the uttermost for his life's set prize," are etched on his headstone at Grytviken—Shackleton believed life was to be lived. That he sought out the majesty of all it could offer and flourished in adversity, almost as if normal life were not worthy of his attentions, was evidenced by the expeditions that he put ahead of all else. His declaration to Worsley, "Never for me the lowered banner, never the last endeavour," really does sum him up. His was the constant quest for challenges and avoiding what he saw as the humdrum day-to-day unless it served a greater purpose. It was the same sort of quality that saw Churchill at his best in wartime—and one that made the politician's private skepticism about the merits of polar expeditions all the more surprising.

Shackleton's love of bold undertakings and sense of greater purpose—along with the fact he'd played only a minor role in the war, having arrived home too late and too old at forty-three—meant that by 1920 Shackleton was yearning again for the great white South. In 1922 he set off on board the *Quest* on an expedition with somewhat ill-defined goals. He traveled with seven men from the *Endurance*, perhaps all seeking to reacquaint themselves with something deeper and more meaningful that, once glimpsed, had proved impossible to forget. Shackleton died of a heart attack on the night of their arrival at South

Georgia, back on the stage where he had enjoyed his finest moments. It was a final act of theater that he would surely have appreciated. I reread the words he uttered to Worsley years before at King Haakon Bay referring to what fate might hold in store for them, obviously unaware of their poignancy: "Some day, Skipper, you and I will come and dig here for old treasure, or perhaps sleep quietly with the other old seamen." Fulfilling his own prophecy, he died on board the *Quest* in Grytviken harbor, aged just forty-seven. Unable to imagine her husband at peace in a cemetery in the gentle pastures of England, Shackleton's widow, Emily, instructed that he should be laid to rest in the graveyard on South Georgia, his grave alone pointing south toward his spiritual home.

We walked up to the whalers' cemetery in Grytviken, where Zaz, our patron, awaited us. The Southern Ocean's weather had, incredibly, brought us all together on the same day in this remotest of places, though we arrived on different vessels with itineraries that should not have allowed us to meet. It followed the wonderful synchronicity that saw us leave civilization on the anniversary of Shackleton's death on January 5 and depart from South Georgia to return to civilization the following day on February 15—his birthday. The power of such timing would have been manna for his sense of the dramatic, our expedition being a story of his death to rebirth, a metaphor for the odyssey he undertook that arose from the wreck of the *Endurance* to eclipse the original journey. That we had managed to retrace this journey as close as it is possible to do in the modern era was humbling. We now realized more than can be expressed in the words we'd all read a hundred times how much they had suffered to achieve what they did and that it was a journey that could never truly be repeated.

Many words have been written about Shackleton over the years but perhaps there are none better than those of Apsley Cherry-Garrard, a survivor of Scott's Terra Nova expedition. "For speed across the ice give me Amundsen, for scientific research there is Scott, but in times of trouble pray God for Shackleton."

Top: The final farewell: Shackleton embarking on the Quest *just six years after his greatest achievement, unaware of the fate that awaited him.*

Bottom: Toasting the Boss: raising our glasses, or rather mugs, at Shackleton's grave.

Over leaves: ". . . heard the text that Nature renders . . ."

SAVING ANTARCTICA FROM MAN

Our expedition provided an opportunity to observe man's impacts on the western part of Antarctica and the sub-Antarctic over the years, commencing with the exploitation of marine resources in the nineteenth and twentieth centuries and, more recently, the effects of climate change.

Exploitation of seals and whales began shortly after Captain Cook's second expedition aboard the HMS *Resolution* in 1775, when he discovered South Georgia. The early polar explorer James Weddell stated in 1825 that "the number of skins brought from off South Georgia cannot be estimated at fewer than 1,200,000." King George Island was a center of sealing until the end of the nineteenth century, by which time fur seals had been hunted to extinction on the South Shetland Islands, with sealers being replaced by whalers. Whalers needed large, deep bays to capture whales, and Admiralty Bay was one of the best. In the period 1911–30 in the South Orkneys and South Shetlands area, the total number of whales taken was 118,159. Today there are still bones from hundreds of whales strewn on the shores of Admiralty Bay from this period. When Shackleton reached Stromness whaling station in 1916, land-based whaling was still at its height, but as little as a decade later whaling on the high seas—"pelagic whaling"—was taking over owing to the scarcity of whales around South Georgia Island caused by overharvesting. Whale-catching ships with a range of around 300 kilometers harpooned whales using an explosive grenade inflated with air, marking the whale for subsequent retrieval. They then towed the harpooned whales to a factory ship or shore station, where the blubber was removed and boiled under pressure to extract the oil. Between 1904 and 1965 some 175,250 whales were processed at South Georgia shore stations. In the whole Antarctic region some 1,432,862 animals were taken between 1904 and 1978, when hunting of the larger species ceased. Probably the largest whale ever recorded was taken at South Georgia; a blue whale processed

at Grytviken in about 1912, it was 33.58 meters (110 feet) long and weighed in at just under 200 tonnes.

In 1986 all commercial whaling was stopped, although some nations such as Japan continue to hunt certain whales for "scientific purposes." Seal numbers are now back to their pre-exploitation levels, but blue and humpback whales are at only a fraction of their prewhaling numbers: 1 percent for blue whales and between 2 and 20 percent for humpbacks.

Man's main impact on Antarctica today is climate change, with global mean warming 0.8°C above preindustrial levels. Oceans have warmed by 0.09°C, while sea levels have risen by about 20 centimeters since preindustrial times and are now rising at 3.2 centimeters per decade.

These changes relate to the accumulation in the atmosphere of excess greenhouse gases, which at the time of writing have cumulatively reached 450 parts per million (ppm) of CO_2 equivalent concentration in the atmosphere. This represents an increase since the industrial revolution of 37 percent for carbon dioxide, 150 percent for methane, and 18 percent for nitrogen oxide.

In terms of Antarctic ice melt, the warming of the atmosphere and oceans is leading to an accelerating loss of ice from the Greenland and Antarctic ice sheets, and this melting is likely to add substantially to sea-level rises in the future. Overall, the rate of loss of ice has more than tripled since the 1993–2003 period as reported by the International Panel on Climate Change, reaching 1.3 centimeters per decade between 2004 and 2008; the 2009 rate is equivalent to a loss of about 1.7 centimeters per decade. If ice loss continues at these rates, the increase in global average sea level due to this source alone would be about 15 centimeters by 2100.

The main part of Antarctica where ice loss is occurring is the West Antarctic Ice Sheet (WAIS), which warmed by about 2.4°C between 1958 and 2010, making it one of the fastest-warming areas of the planet. The WAIS, to the west of the Transantarctic Mountains, including the Antarctic Peninsula, makes up about 10 percent of the total volume of ice in Antarctica, and alone contains enough ice to raise sea levels by at least 3.3 meters.

Data for ice melt on Elephant Island is not readily available, but a great deal of data exists for other islands in the South Shetland Island group, in particular for King George Island (only 80 nautical miles from Elephant Island), including Arctowski base, where the team spent almost two weeks sea-trialing the *Alexandra Shackleton* and undertaking crevasse training. King George Island provides an excellent test case for predicting future responses to climate change as it is undergoing dramatic glacier recession and environmental change.

Arctowski is situated in Admiralty Bay and was established in 1977. In terms of sea ice, the sea in Admiralty Bay froze completely eleven times between 1977 and 1996 during the winter. In the period 1999–2012, it did not apparently freeze once.

As a direct consequence of climate change, reduced sea ice in Admiralty Bay has unquestionably contributed (along with pollution and human disturbance) to massive reductions in penguin numbers in the colonies adjacent to Arctowski. Adélie penguins, who breed on ice-free land but live on sea ice, have, for example, declined from approximately 10,000 pairs in 1978 to only 3,600 pairs in 2012. The Chinstrap population at Arctowski, meanwhile, is down from 500 breeding pairs in 1978 to one breeding pair today. Although Chinstraps prefer to live in ice-free sea, their main food source, krill, depends on algae that attaches to the underside of sea ice. Reduced sea ice means less algae, which in turn means less krill and fewer Chinstrap penguins. The third type of penguin present at Arctowski, Gentoo penguins, are also down from 780 breeding pairs to 70 pairs; again, a major contributor is the warming of the sea around King George Island and the consequent reduction in krill quantities. Interestingly, a fourth type of penguin was observed by the expedition team at Arctowski in January 2013—a solitary King penguin. Normally a sub-Antarctic penguin, Kings are now occasionally seen farther south than they would otherwise be, perhaps an indication that the climate is increasingly suitable for them.

The Windy, Baranowski, and Ecology glaciers next to Arctowski base are part of the Warszawa ice field, which is part of the King George Island ice cap. During the past decade, a rapid retreat of these glaciers has occurred, consistent with warmer temperatures in the region. The Ecology Glacier has been receding continuously since at least 1956–57, with rates ranging from 4 meters per year in the late 1980s to up to 30 meters a year in the decade 1989–1999. The Windy and Baranowski glaciers reveal similar results.

The expedition's final destination was South Georgia, over half of which is covered in "permanent" snow and ice. Based on historical maps, satellite images, and other data, the history of South Georgia's glaciers can be traced. A study in 2010 of 103 of the 160 glaciers on the island from the 1950s to the present found that 97 percent retreated during this period. At low elevations, some are even approaching disappearance.

Our expedition team certainly felt this to be the case. When Shackleton crossed South Georgia, he encountered three large glaciers: the Crean, Fortuna, and König. Our team followed the same route and found that the Fortuna appeared largely intact, while the Crean was very heavily crevassed, with virtually no snow covering the crevasse mouths. This is partly attributable to the fact that we visited South Georgia at a warmer time of year, but it also relates to the effects of warming and increased precipitation. Meanwhile, whereas Shackleton had to cross the end of the König Glacier, our modern team found a grassed alpine meadow, the König having retreated many kilometers up the valley.

Temperature and precipitation levels since 1930 have both been rising, which generally leads to smaller glaciers. The South Georgia glacier changes are broadly matched by those on other sub-Antarctic islands, including Heard Island, Kerguelen, and the South Shetlands including Elephant Island, as well as in southern South America.

Another interesting indicator of climate change on South Georgia witnessed by the expedition team is the reindeer eradication program currently being undertaken by the South Georgia government. Reindeer, which were introduced to South Georgia by the Norwegians in the 1900s as a food source, are now able to venture farther inland as the natural physical barriers provided by the glaciers have retreated. Reindeer eradication is occurring because the flora and fauna in the island's interior cannot cope with grazing and trampling—a problem the South Georgia government knows will only increase with the ongoing climate change that is sadly going to occur.

ACKNOWLEDGMENTS

The majority of an iceberg remains hidden from view, and so it is with this expedition. Six of us undertook the boat journey, while three of us took the final steps into Stromness after a grueling journey across South Georgia's mountains, but we are part of a much larger team, all of whom played their part in our achievement. To you all, I say our success is your success. Of particular note:

The Honorable Alexandra Shackleton (Zaz), for asking me to lead the expedition and for providing her patronage and unstinting support throughout.

The expedition team: Nick Bubb (skipper), Paul Larsen (navigator), Seb Coulthard (bosun), Barry Gray (mountain leader), Ed Wardle (cameraman), Paul Swain (reserve sailor). A more capable team of men would be difficult to find. They made the world's most difficult survival journey seem manageable.

Sponsors, without whom none of this would have been possible: Naming rights sponsor Intrepid Travel and in particular Geoff "Manch" Manchester, Robyn Nixon, and Eliza Anderson for their tremendous support and warmth. Major sponsors St. George Bank: special thanks to Andy Fell for his "can-do" approach. Whyte & Mackay and their Mackinlay's Highland Malt based on Shackleton's original: to Richard Patterson, Rob Bruce, Jill Inglis, and Chris Watt—"Slange Var."

My employers and major sponsors Arup, who are wonderful supporters of my endeavors, and in particular Peter Bailey, Robert Care, Philip Dilley, John Clay, James Kenny, Miles King, Martin Ansley-Young, Julian Brignell, Ramona Dalton, Ben Richardson, and Piers van Till.

The James Caird Society, in particular Sir James Perowne KB, Stephen Scott-Fawcett, David Tatham, Pippa Hare, Dorothy Wright, Nick Smith, and, of course, Zaz, for their moral and financial support. To John Leece/Borough Mazars, Bobby Haas, Dick Smith, Dr. Martin L. Greene, and James Caird Asset Management for their generous donations that made such a big difference to things.

Kim McKay and her wonderful team at Momentum 2, Biarta Parnham, Jo Stewart, and Simone Bird, who helped with all aspects of media, PR, and fundraising for the project and continue to be great supporters, and who arranged for Emanate PR to assist us with our media launches in both New York (Marissa Mastellone) and London (Lishai Kaufer).

Aurora Expeditions, who provided free transport for the Alexandra Shackleton to and from Antarctica on board Polar Pioneer. A special thanks to Lisa Bolton, CEO of Aurora, Captain Yury Gorodnik, skipper of MV Polar Pioneer, and the outstanding polar logistician Tomas Holik, who worked tirelessly to help us.

Our conservation partners and the oldest and surely the most effective environmental charity in the world, Fauna & Flora International, in particular CEO Mark Rose, Joe Heffernan, Ally Catterick, and Clare Verberne. I look forward to working together.

The sponsor team who accompanied us as far as Arctowski and who were a great support to us: Julian Brignell, Ashley Henley, Keith Hewett, Steve Lennon, and Donald Ewen from Arup, Intrepid Travel's Jane Crouch, Sam Tomaras from St. George Bank, and Virgin Media/Discovery Channel competition winner Nigel Sinclair.

Nat and Gill Wilson and all at the International Boatbuilding Training College, who built the Alexandra Shackleton for the love, one that got us safely across the roughest ocean in the world in one piece—the warmest of wishes.

All at Discovery Channel and PBS for having confidence in the program and supporting it wholeheartedly, and to NBC for promoting it in conjunction with PBS.

To the crew of Australis—skipper Ben Wallis, Magnus O'Grady, and Skye Marr-Whelan—for their fantastic judgment and "can-do" attitude.

Margot Morrell for her help, support, and friendship in researching all aspects of Sir Ernest Shackleton and his team and who surely knows more about him than anyone bar Zaz.

John Quigley of QXI for helping us insure the uninsurable.

For wonderful legal and contractual advice provided pro bono by Sydney's leading law firm Corrs Chambers Westgarth, in particular Trevor Danos (partner), Eugenia Kolivos (partner), and Heather Hong (lawyer).

"The Raw crew"—Jamie Berry (producer), Joe French, and Si Wagen (cameramen)—for their skill and expertise in filming this epic journey, and Sam Maynard, Piers Vellacott, Ben Barrett, Kathryn Taylor, Marie Frellesen, Nageena Ahmed, and Georgia Woolley for assisting with all other aspects of making this film happen.

The Foreign & Commonwealth Office, in particular Henry Burgess, Deputy Head, Polar Regions Unit; and Oscar Castillo, Desk Officer, South Georgia and the South Sandwich Islands; and the government of South Georgia and the South Sandwich Islands, in particular Richard McKee, executive officer, Caradoc "Crag" Jones, and government officer Pat Lurcock.

Our friends at Arctowski base: Marek Zieliżski, Jacek Czarnul, Radosław Łabno, Sylwia Łukawska, Kazimierz Połeż, Waldemar Swołek, Włodzimierz Tkaczyk, Agata Wilk, Piotr Spiczyn—"Najlepsze Życzenia!"

All those organizations who helped us by providing expertise, training, and gear and equipment, including Sunspel clothing, the Royal Navy, the Royal Marines, Dean & Reddfyhoff Marinas, Ocean Safety, Yellowbrick Tracking, Davey & Company, Henri Lloyd, Suffolk Barrel, Bussells, Compass Marine, Medic Services International, Revive Portland, UK Sail Training, Trinity Sailing, Morning Star UK, Wolfson Unit MIA, the RNLI, the Scott Polar Research Institute, DHL Poland, and Dulwich College.

Friends, supporters, and advisers who have helped make the project a reality and variously provided funds, support, their time, advice, and goodwill, or all of the above. Particular thanks to legendary sailor Trevor Potts; the UK's leading traditional sailmaker, Philip Rose-Taylor, who handcrafted our sails so expertly; logistics expert Howard Whelan; Georgia Woolley, Nageena Ahmed, and Ed Wardle for sourcing period gear; Skip Novak for his polar advice and know-how; Graham Neilson, Pat and Sarah Lurcock for such a warm welcome on South Georgia; Dave Stace for web magic; Julian "Woody" Woodall for medical advice; Iain Kerr and John Kirkwood for their understanding; Andrew "Ed" Edwards and DHL for boat transport; Patrick and Gigi Blumer, Mary and Anthony Blumer, Robert Blumer, Jean and John Jarvis, Dan Jarvis for their love and support; Sophie Bubb, Helena Darvelid, Joanne Gray (editor of *BOSS Magazine* at *Australian Financial Review*); Mike Rann, kindred spirit Cheryl Bart, chair of the South Australian Film Corporation, and Greg Marsh, business affairs manager, Captain Mike Forwood, Calista Lucy, Melissa Shackleton-Dann and Tom Dann, Richard Dennison, head of Orana Films, for his love of the story and friendship; Nicolas Paulsen, DAP Airlines; Jonathan Shackleton; Professor Mike Tipton for teaching us the consequences of falling in; Alan Rowe MBE (founder of The Baton charity); Commander Louis Wilson-Chalon, CO 815 Naval Air Squadron (2010–2013), Commander Alastair Haigh, CO 815 Naval Air Squadron (2013–present), Lieutenant Commander Philip Richardson RN, Senior Pilot, 815 NAS, Lieutenant James Beedle RN and ship's company of HMS *Blazer*, Chief Petty Officer Clive Russell RN, Chief Petty Officer Paul Mclean RN, and Color Sergeant Pete Wooldridge RM.

And to all those who helped us make the *Alexandra Shackleton* fit to go to sea, including testing, sea trials, and training: Barry Deakin, Able Seaman Eddie Janion, Able Seaman John Rushton, James Harriott, Dr. Robert Goodhart, who worked tirelessly to help us in so many ways, Ian Baird, Philip Ambler, Fiona Lewis, Yvonne Beven (a home away from home), Paul Swain, Russ Levett, Roly Gill, Nick Coutts, David Thomas, Scott Irvine (photographer), Dave and Jackie Baker, Captain Bob Turner RN (a former captain of the ice patrol ship HMS *Endurance*), marine electrician Robert Sleep Mike Winter, Barry Alborough, Vicky Farrar, Robin Stone, Hannah Leach, John Dean and Richard Reddyhoff, Chris Waterman, Alistair Hackett, Erica Mear, Nick Farrell, Peter Tracey, Ian Howieson, Fred Attwood, Trevor Gray, Chris Stanmore-Major, Julian Woodall, Mark Fagg, and, at the *Alexandra Shackleton*'s future home at Portsmouth Historic Dockyard, Boathouse No. 4, Jim Brooke-Jones.

To all at HarperCollins, in particular Amruta Slee (publisher), Julia Collingwood and Matt Stanton and Myles Archibald (publishing director, UK), for all their patience and hard work in helping produce what I hope will be a great book and worthy of the film it accompanies.

Last but not least, I would like to thank Elizabeth, who has loved and is loved and who has been a huge support to me both before, during, and since the expedition. I'm lucky to have her.

***

There are many individuals and organizations to whom I am grateful for allowing me access to the words of others.

All extracts from *Shackleton's Boat Journey* by F. A. Worsley, published by Pimlico, are reprinted by permission of the Random House Group Limited in London and W. W. Norton in New York.

I am grateful to Alexandra Shackleton (Zaz) for authorizing the use of quotations from Sir Ernest Shackleton's diaries, papers, and correspondence as well as those from the correspondence of Reginald James, all of which also appear by permission of the University of Cambridge, Scott Polar Research Institute.

Excerpts from the diaries of Alexander Macklin and Thomas Orde-Lees also appear by permission of the University of Cambridge, Scott Polar Research Institute, as does the letter dated January 14, 1914, to Shackleton from Misses Pegrine, Davey, and Webster.

The families of Alexander Macklin and Lionel Greenstreet were kind enough to give their authorization for the use of extracts from the diaries of these two *Endurance* crew members. The Mitchell Library, State Library of New South Wales, gave permission for me to quote from the ITAE journal of Frank Hurley (MLMSS 389 Boxes 2 and 5) as well as the extract from Frank Wild's memoirs (MLMSS 2198 Vol. 2). Similarly, the Alexander Turnbull Library in Wellington, New Zealand, granted permission for the use of the excerpt from the Journal of Harry McNeish (MS-1389) as well as the excerpts from the Antarctic Journal of Thomas Orde-Lees (1877–1958) (qMS-1149-1153).

I am also grateful to Jamie Young of the South Aris team for generously allowing us to quote his words from his letter to Arved Fuchs about his experiences in the Southern Ocean. Similarly, I am grateful to Peter Hillary for giving me his permission to cite his father Sir Edmund Hillary, including his version of Sir Raymond Priestley's words about Shackleton and his heroic-era contemporaries.

The excerpt from William Pollock's article in the *Daily Mirror* is reproduced courtesy of the *Daily Mirror*/Mirrorpix.

Editions Gallimard in Paris kindly granted permission to quote the words of André Gide, taken from *Journal 1887–1925* (Copyright © Editions Gallimard, Paris, 1996). The words of fictional sage Professor Dumbledore from *Harry Potter and the Chamber of Secrets*, copyright © J. K. Rowling 1998, are reproduced with the kind permission of J. K. Rowling.

The lines from "The Waste Land" are taken from *Collection*, copyright © Estate of T. S. Eliot and are reprinted by permission of Faber and Faber Ltd, to whom I am also grateful for permission to use lines from "Thalassa," from *Collected Poems* by Louis MacNeice, also published by Faber and Faber.

Churchill's famous words from his speech at the Lord Mayor's Luncheon at Mansion House on November 10, 1942 (copyright © Winston S. Churchill), are reproduced with permission of Curtis Brown, London, on behalf of the Estate of Sir Winston Churchill. Those of Helen Keller, © the American Foundation for the Blind, Helen Keller Archives, are reproduced with the permission of the American Foundation for the Blind.

PICTURE CREDITS